CW00982903

Powering the Green Economy

Powering the Green Economy
The Feed-in Tariff Handbook

*Miguel Mendonça, David Jacobs
and Benjamin Sovacool*

London • Sterling, VA

First published by Earthscan in the UK and USA in 2010

Copyright © World Future Council, Benjamin K. Sovacool and David Jacobs, 2010

All rights reserved

ISBN: 978-1-84407-857-8   hardback
        978-1-84407-858-5   paperback

Typeset by JS Typesetting Ltd, Porthcawl, Mid Glamorgan
Cover design by Yvonne Booth

For a full list of publications please contact:

Earthscan
Dunstan House
14a St Cross St
London, EC1N 8XA, UK
Tel: +44 (0)20 7841 1930
Fax: +44 (0)20 7242 1474
Email: earthinfo@earthscan.co.uk
Web: **www.earthscan.co.uk**

22883 Quicksilver Drive, Sterling, VA 20166-2012, USA

Earthscan publishes in association with the International Institute for Environment and Development

A catalogue record for this book is available from the British Library

Library of Congress Cataloging-in-Publication Data
Mendonça, Miguel.
    Powering the green economy : the feed-in tariff handbook / Miguel Mendonça, David Jacobs, and Benjamin Sovacool.
        p. cm.
    Includes bibliographical references and index.
    ISBN 978-1-84407-857-8 (hardback) – ISBN 978-1-84407-858-5 (pbk.) 1. Energy policy–Environmental aspects. 2. Energy industries–Prices. 3. Renewable energy sources–Government policy. 4. Energy development–Environmental aspects. 5. Energy development–Government policy. I. Jacobs, David, 1978- II. Sovacool, Benjamin K. III. Title.
    HD9502.A2M463 2009
    333.79'4–dc22
                                                    2009024721

At Earthscan we strive to minimize our environmental impacts and carbon footprint through reducing waste, recycling and offsetting our CO_2 emissions, including those created through publication of this book. For more details of our environmental policy, see www.earthscan.co.uk.

This book was printed in the UK by TJ International, an ISO 14001 accredited company. The paper used is FSC certified and the inks are vegetable based.

To Louise and Joe, and in memory of Dexter Banks.
–MBM

To Lutz Mez and Danyel Reiche for their encouragement in continuing my research on renewable energy support mechanisms; and Fabian for the inspiring conversation in Amsterdam.
–DJ

To Richard Hirsh: without you, I never would have become interested in renewable energy or, more important,
energy efficiency.
–BKS

We are now faced with the fact that tomorrow is today. We are confronted with the fierce urgency of now. In this unfolding conundrum of life and history there is such a thing as being too late. Procrastination is still the thief of time. Life often leaves us standing bare, naked and dejected in lost opportunity. The 'tide in the affairs of men' does not remain at the flood; it ebbs. We may cry out desperately for time to pause in her passage, but time is deaf to every plea and rushes on. Over the bleached bones and jumbled residue of numerous civilizations are written the pathetic words: 'too late.'

Martin Luther King, Jr., 1967

Hurry up please, it's time.
Hurry up please, it's time.

T. S. Eliot: The Waste Land, 1922

It's now or never.

Elvis Presley, 1960

Contents

Figures and tables	*ix*
Forewords	*xiii*
Acknowledgements	*xvii*
Acronyms and abbreviations	*xix*
Introduction	*xxi*
1 The Green Economy	1
2 Basic FIT Design Options	15
3 Advanced FIT Design Options	39
4 Bad FIT Design	57
5 FIT Design Options for Emerging Economies	67
6 FIT Developments in Selected Countries	77
7 Dispelling the Myths about Technical Issues	111
8 Barriers to Renewable Energy Deployment	129
9 Other Support Schemes	149
10 Campaigning for FITs	185
Notes	*195*
Index	*199*

Figures and Tables

FIGURES

0.1	A simplified diagram of how a typical FIT changes the supply and demand of electricity	xxii
0.2	The Industrial Revolution 2.0	xxvi
1.1	Expected investments (left column) versus investment potential (right column) for renewable electricity systems around the world, 2004–2030	5
2.1	German methodology and input variables for calculating electricity production costs	22
2.2	General flow of electricity and financing under FIT schemes	29
2.3	Percentage of grid integration costs compared to total investments	31
2.4	Cost-sharing methodology for grid connection	32
3.1	Progression of the remuneration level under the Spanish premium FIT (wind onshore)	41
5.1	Feed-in tariff fund for developing countries	70
5.2	Mini-grid powered with renewable energy sources	73
5.3	Feed-in tariff for mini-grids (IPP)	74
5.4	Feed-in tariff for mini-grids (producer/consumer)	75
5.5	Feed-in tariff for mini-grids (RESCO)	75
6.1	Diffusion of support mechanisms in the EU	79
6.2	Share of renewable energies in the German gross electricity consumption	81
6.3	The merit-order effect	84
6.4	Locations in the US with FIT legislation and/or regulatory initiatives (as of May, 2009)	94
7.1	The 23.2MW *Kombikraftwerk*, a hybrid wind–solar–biogas–hydro facility serving Schwäbisch Hall, Germany	119
7.2	The ultra-efficient Rocky Mountain Institute headquarters in Snowmass, Colorado	120
7.3	A typical pumped hydro storage system	122

7.4	A typical compressed air energy storage system	123
7.5	A typical molten salt storage facility	124
8.1	Annual growth of installed wind power capacity in Denmark	135
8.2	Energy research subsidies in OECD countries, 1974–2002	137
8.3	Annual bird deaths in Denmark and Britain caused by wind turbines, cars and cats	143
8.4	The 'external costs' of conventional, nuclear and renewable power generators in the US	143
8.5	The 'external costs' of power generators compared to their production costs in the US	144
9.1	Renewable energy capacity built in states with and without renewable portfolio standards (RPS) in the US, 1978–2006	154
9.2	Price volatility of RECs in the US, 2003–2008	159
9.3	Green power programmes in the US, 2008	161
9.4	Simplified diagram of net metering showing power flows to/from the grid	165
9.5	Global energy R&D expenditures, 1974–2007	168
9.6	Global energy R&D expenditures on renewable energy, 1974–2007	168
9.7	System benefits charges and revenues in the US, March 2008	169

TABLES

0.1	The economic, political, social and environmental advantages of FITs	xxvii
0.2	The economic, political, social and environmental challenges facing FITs	xxviii
1.1	Clean energy details of American Recovery and Reinvestment Act	3
1.2	Economic stimulus packages and green portions	4
1.3	Top five renewable energy nations: selected indicators	9
2.1	Summary of basic data and parameters used for profitability calculation	21
2.2	South African methodology for calculating levelized costs	25
3.1	Demand-oriented tariff payment under the Hungarian FIT scheme	43
3.2	Location-specific remuneration for onshore wind power under the French FIT scheme	48
3.3	Tariff degression rates under the German FIT scheme	49
3.4	Flexible degression under the German FIT scheme	51
3.5	Flexible tariff degression as proposed by the industry association ASIF (Spain)	51
3.6	Remuneration for solar PV under the Italian FIT scheme	54

6.1	Feed-in tariffs worldwide as of April 2009	78
6.2	Remunerations under the German feed-in tariff scheme	82
6.3	Costs and benefits of the German FIT scheme	85
6.4	Tariff payment under the Spanish feed-in tariff scheme	86
6.5	The Ontario Power Authority's revised FITs, May 2009	92
6.6	Expression of interest in Ontario Power Authority's FITs	92
6.7	Vermont's recently enacted FITs	95
6.8	Australian Capital Territory feed-in tariff details	98
6.9	New South Wales feed-in tariff details	98
6.10	Northern Territory feed-in tariff details	99
6.11	Queensland feed-in tariff details	99
6.12	South Australia feed-in tariff details	100
6.13	Tasmania feed-in tariff details	100
6.14	Victoria premium feed-in tariff details	101
6.15	Victoria standard feed-in tariff details	101
6.16	Western Australia feed-in tariff details	102
6.17	Tariff payment under the Kenyan feed-in tariff scheme	103
6.18	The energy cost associated with using medium-speed diesel plants in Mombasa and Nairobi (August 2007)	104
6.19	Tariff payment under the South African feed-in tariff scheme	105
8.1	Avian mortality for fossil fuel, nuclear and wind power plants in the US	142
9.1	Countries, provinces and states with renewable portfolio standards and quota schemes	151
9.2	Number of utility green power participants for the ten most successful programmes, December 2008	163

Forewords

I

Feed-in tariffs (FITs) have proven to be the best support mechanism to rapidly increase the share of renewable energy production and use. In Germany, the Renewable Energy Sources Act (EEG) managed to increase the proportion of renewable electricity from 6 per cent in 2000 to more than 15 per cent in 2008. The feed-in tariff scheme has become the most important climate protection tool, far more effective than the EU's Emissions Trading Scheme. In Germany alone, the feed-in tariff has saved the emission of 79 million tonnes of carbon dioxide equivalent (CO_2e) in 2008, an amount greater than the annual emissions of Armenia, Botswana, Cambodia, Cameroon, El Salvador, Iceland, Ireland, Paraguay and Senegal combined.

Last year, the German Green Party decided that by 2030 all electricity shall be provided by renewable energy. This is an ambitious target but an absolute necessity in the light of global climate change and energy security. When we decided in 2000 that we wanted to double the share of renewable electricity by 2010, many people said we were fools and argued that this was impossible. Thanks to the feed-in tariff scheme, we had already doubled the share by 2007. Today, it is not enough to reduce our greenhouse gas emissions. We will have to stop emitting dangerous greenhouse gases, and this is only possible by completely shifting to a renewables-based energy system.

This fundamental change in our energy supply offers new possibilities for society as a whole. In Germany, 30,000 people were working in the field of renewable energies in 1998. A decade later it is almost 300,000 and soon it will be 500,000. The German industry has a turnover of €30 billion, of which a large share is due to technology exports. As feed-in tariffs are independent from governmental spending they have proven to be an effective tool in overcoming the current economic and financial crisis, in contrast to stimulus packages all over the world which are financed from tax revenues. Feed-in tariffs therefore provide a stable and successful incentive for new investment and job creation, without public borrowing.

Even though feed-in tariffs are the best way to support renewable energies, the actual design of this instrument is crucial for cost-efficient and effective support. Here is where the *Feed-in Tariff Handbook* becomes instrumental. In past years, we have seen many governments from different parts of the world considering or implementing feed-in tariffs, including North America, China, South Africa and Australia. This book is an essential tool to help policy makers all over the world to design well-functioning feed-in tariff schemes, and thus contribute to protecting our climate, securing our energy needs and enhancing international peace. I highly recommend learning from it.

Hans-Josef Fell, MP

Hans-Josef Fell is Member of the German Parliament and author of the proposal of the German feed-in tariff scheme of 2000 (the EEG). He is Vice-President of EUROSOLAR and a member of the World Council for Renewable Energy.

II

Levelling the playing field – the benefits of feed-in tariffs

French scientist Alexandre Edmond Becquerel first discovered the photovoltaic effect in 1839. Albert Einstein won the Nobel Prize in 1921 for explaining it. The effect was first put to work in the US satellite Vanguard I in the early 1950s. But it has only been put to broad commercial use in the new millennium.

The history of photovoltaic technology shows that the key to its successful proliferation lies as much in the sphere of public policy as in the laboratory. It was not until governments began promoting demand for renewable energy – particularly in Europe – that private investors were willing to make significant investments in the mass production of solar panels. This, in turn, has led to sharply lower costs and dramatic innovation across the entire photovoltaic value chain, demonstrating in just a few years the enormous potential of photovoltaics (PV) in the low-carbon energy infrastructure needed to secure our future.

German policy makers including Alternative Nobel Prize winner Hermann Scheer and Green Party Energy Spokesman Hans-Josef Fell had a clear vision in the beginning of 2000 that only a reliable mass market would allow PV manufacturers to achieve economies of scale that could reduce the cost of photovoltaic electricity over time. The German Renewable Energy Act (EEG) guarantees that a 'feed-in tariff' is paid – at a fixed rate for 20 years – for each kilowatt-hour generated by renewable energy. The tariff allows every participant along the entire value chain to make a reasonable profit, on the expectation that this profit will be reinvested in further scaling up production and reducing costs.

The feed-in tariff has greatly accelerated the progress of First Solar and other solar companies in readying photovoltaics for mass market adoption. In 2005, our first full year of production, we produced 20MW of solar modules. We entered 2005 with manufacturing costs of US$3 per watt. The provisions of Germany's 2004 EEG provided us the market we needed to attract expansion capital and scale up our operations quickly and cost-effectively. While scaling up production in our first plant in Perrysburg, Ohio, we also invested €115 million in a 100MW production facility in eastern Germany. In 2009, we expect to have a global production capacity equivalent to 1GW of peak power at production costs below $1 per watt – including an innovative, pre-funded recycling scheme for modules that reach the end of their lives. The EEG helped us become not just one of the largest PV manufacturers in the world but also one of the most sustainable manufacturers of any type.

This success would have been impossible without the introduction of the feed-in tariff in Germany and other countries. More than 40 states, countries or regions in the world have adopted some version of the feed-in tariff. Not all of them have been as successful as Germany, but all of them have enabled much quicker, more affordable and more sustainable growth in clean energy industries than alternative support schemes.

Down the road there is no alternative to achieving a low-carbon energy infrastructure. This book shows the path. The sooner feed-in tariffs are adopted and expanded in markets worldwide, the sooner we can achieve the 100 per cent objective. I congratulate the authors for having put together such a useful and insightful manual for legislators and other decision makers.

Mike Ahearn
Chairman, First Solar

Acknowledgements

Miguel Mendonça thanks, for their assistance, input and inspiration: Herbie Girardet and Azad Shivdasani; David Jacobs and Benjamin Sovacool; the staff, funders, advisors and councillors of the World Future Council; the city of Hamburg and Michael Otto; Mike Fell, Claire Lamont, Hamish Ironside and Dan Harding at Earthscan; Achim Steiner, Nick Nuttall and Maxwell Gomera (UNEP); David Suzuki; Frances Moore Lappé, Daniel Kammen, Peter Coyote, Paul Gipe, Jose Etcheverry, Janet Sawin, Bianca Barth, Randy Hayes, Lois Barber and the steering committee of the Alliance for Renewable Energy; Rep. Jay Inslee; Alan Simpson MP, Leonie Green, Dave Timms, David Toke and the entire UK tariff coalition; Stephen Lacey, Lynda O'Malley, Lily Riahi and Brook Riley; Toby Couture, Manu Sankar, Jaideep Malaviya, David Moo, Jonathan Curren, Charmaine Watts and Stefan Gsänger; an innumerable list of thinkers, writers, artists, actors and activists, without whom I would never have written a word; my nearest and dearest: Mike Wallis, Fiona Balkham, Daphne Kourkounaki, Dennis Keogh, Daniel Oliver; the Mendonças and the Blachfords; and especially Lou and Joe.

David Jacobs thanks: The German Federal Foundation for the Environment (DBU) for financing the research on support mechanisms for renewable electricity, Danyel Reiche, Lutz Mez and Miranda Schreurs from the Environmental Policy Research Center (FFU), The World Future Council, Carsten Pfeiffer, Hans-Josef Fell, Fabian Zuber, Thomas Chrometzka, Janet Sawin, Mischa Bechberger, David Wortmann, Florian Valentin, France, Frigiliana, Fritz, Ulla and Mascha.

Benjamin K. Sovacool thanks: The Lee Kuan Yew School of Public Policy and the Centre on Asia and Globalisation for their enthusiastic support of the book and for public policy for renewable energy as a whole. Paul Gipe took the time to provide much-needed information relating to recent events concerning FITs in Canada, and Wilson Rickerson and Toby Couture were instrumental in revising the section on FITs in the US. All three deserve very special thanks. Lastly, Dr Sovacool is grateful to the Singaporean Ministry of Education for grant T208A4109, which has supported elements of the work reported here. Any opinions, findings, and conclusions or recommendations expressed in this material are those of the author and do not necessarily reflect the views of the Centre on Asia and Globalisation or the Singaporean Ministry of Education.

Acronyms and Abbreviations

$/kWh	dollars per kilowatt-hour
€/kWh	euros per kilowatt-hour
ADEME	Agence de l'Environnement et de la Maîtrise de l'Energie
AWCC	Average Weighted Cost of Capital
BEE	German Renewable Energy Federation
BIPV	building-integrated photovoltaics
BMU	German Federal Ministry for the Environment, Nature Conservation and Nuclear Safety
Btu	British thermal unit
CAES	compressed air energy storage
CDM	Clean Development Mechanism
CO_2e	carbon dioxide equivalent
COAG	Council of Australian Governments
CSP	concentrated solar power
DECC	Department of Energy and Climate Change
DSO	distribution system operator
EEG	Renewable Energy Sources Act (Germany)
EU	European Union
EU ETS	European Union Emissions Trading Scheme
FERC	Federal Energy Regulatory Commission (US)
FIT	feed-in tariff
G-20	Group of Twenty Finance Ministers and Central Bank Governors
GATS	Generation Attribute Tracking System
GDP	gross domestic product
GEA	Green Energy and Green Economy Act
GIS	Generation Information System
GO	Guarantee of Origin
GSA	General Services Administration
GW	gigawatt
HUD	US Department of Housing and Urban Development
IEA	International Energy Agency

IfnE	Ingenieurbüro für Neue Energie
IRENA	International Renewable Energy Agency
IPP	independent power producer
IRR	internal rate of return
ITC	investment tax credits
kWh	kilowatt-hour
MNRE	Ministry of New and Renewable Energy
MW	megawatt
MWh	megawatt-hour
NERSA	National Energy Regulator of South Africa
NGO	non-governmental organization
NFFO	Non-Fossil Fuel Obligation
NPV	net present value
OPC	Ontario Power Authority
O&M	operation and maintenance
OECD	Organisation for Economic Co-operation and Development
OPA	Ontario Power Authority
PI	profitability index
PTC	production tax credit
PURPA	Public Utility Regulatory Policies Act
PV	photovoltaic(s)
R&D	research and development
REA	Renewable Energy Association
REC	renewable energy credit
REGOs	Renewable Energy Guarantee of Origin Certificates
RES-c	electricity from renewable energy sources
RESCO	rural energy service company
RO	Renewables Obligation
ROC	Renewables Obligation Certificate
RPO	Renewable Purchase Obligations
RPS	renewable portfolio standards
SBC	system benefits charge
SREC	Solar Renewable Energy Certificate
TGC	tradable green certificate
TSO	transmission system operator
UNEP	United Nations Environment Programme
UNFCCC	United Nations Framework Convention on Climate Change
WHO	World Health Organization

Introduction

This book is about the most effective policy for promoting renewable energy, and thereby creating the fundamental conditions for a green and stable economy. It is called a feed-in tariff, or FIT for short. It is a policy with many names, including REPs (Renewable Energy Payments) and ARTs (Advanced Renewable Tariffs).

FITs set a fixed price for purchases of renewable power, usually paying producers a premium rate over the retail rate for each unit of electricity, or kilowatt-hour (kWh), fed into the grid. FITs usually require power companies to purchase all electricity from eligible producers in their service area at this premium rate, over a long period of time. FITs also often force all electric utilities and transmission operators to connect all possible renewable power providers to the grid, and mandate that the utilities themselves pay the interconnection costs, or at least the grid expansion costs. These costs are then distributed among all electricity consumers, minimizing costs while delivering an ever-growing amount of renewable energy. It may not look like it, but a FIT is a truly revolutionary tool – one that changes the role that governments, power operators, grid operators, transmission and distribution operators, and ordinary consumers currently play when it comes to electricity (See Figure 0.1). As this book will exhaustively document, regulators generally differentiate FITs by technology, plant size, location and time to reflect differing production costs, and there are many other potential design options – not all of them good ones.

When designed and implemented properly, however, FITs can bring great advantages to electricity consumers, electric utilities, politicians, businesses, farmers and society at large. FITs are not, as has been argued against them, simply a way for rich people who can afford solar panels to make other people pay for them. They are a way for consumers wishing to generate their own power to receive guaranteed payments, and benefit from additional revenue and the improved reliability of energy supply. These benefits spill over and help all consumers by lowering electricity prices. Electric utilities benefit from displaced fuel costs and decreased volatility of fuel and electricity prices. Politicians benefit because FITs often jump-start a robust manufacturing sector for renewable electricity technologies, bringing with them tax revenue and high-paying jobs that stay within the community.

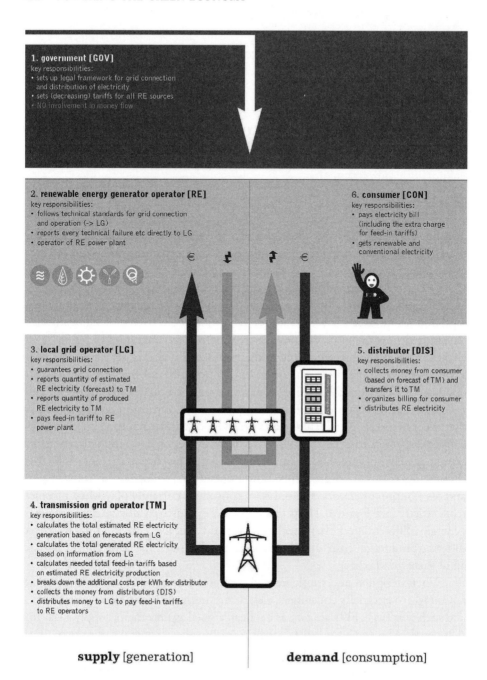

Figure 0.1 *A simplified diagram of how a typical FIT changes the supply and demand of electricity*

Source: European Renewable Energy Council, 2007

Businesses and farmers, among other groups, can install generation equipment and gain extra income, and society benefits from reduced greenhouse gas emissions and greater diversification of the electricity sector. Properly designed FITs can deliver all these things, at low cost.

Why do we need more well-designed FITs? The Intergovernmental Panel on Climate Change tells us that we have less than a decade to meaningfully start reducing greenhouse gas emissions. All experts in this field agree that renewable energy – and energy efficiency – will play a key role in this battle. FITs help spur technological development through rapid deployment and economies of scale, thus decreasing generation costs of renewable energy sources and improving their competitiveness compared to conventional electricity systems based on coal, gas, oil and nuclear. The greater the number of countries implementing well-designed FIT schemes today, the sooner the costs of renewable electricity will fall below the price of conventionally produced electricity. This is no longer a question of decades, but of years. Once we have reached this tipping point, FITs will have done their job, and will only be needed on a limited basis, if at all.

Examples from all over the world show that renewable energy technologies have the potential to become the backbone of our energy system. The costs of wind energy and several other renewable energy technologies are already within the range of prices paid on electricity spot markets in many countries. The cost for solar photovoltaics (PV), long thought to be the most expensive renewable resource on the market, should reach 'grid parity', i.e. the price that final consumers pay for the electricity, within the next decade. In developing countries, most renewable energy technologies are more cost-effective than expensive electricity from diesel generators. By implementing FITs, the remaining gap between the increasing costs of fossil fuels and the decreasing costs of renewable energy will be closed faster.

A robust discussion about FITs is therefore instrumental given the energy and resource challenges our societies currently face. The kinds of transition necessary in our human systems can, by their nature, bring us back to a cyclical way of life, rather than our man-made linear flows from sources to sinks. Nature is fantastically adept at recycling, and living within natural limits. As described by many ecological thinkers of the last half century in particular, it is a fundamental truth that humans are part of nature too. We arguably organize, interact and manipulate materials in more sophisticated ways, but essentially we are self-propelled biological systems dependent upon natural inputs. Energy systems based on renewable sources are a partial attempt to replicate this, and also help to attune people to our connection to nature, and demonstrate the ability to exploit it almost harmlessly. The level of environmental damage created by using renewables is negligible compared to the processes, supply chains, pollutants, political corruption, terror threats, climate change effects and other externalities involved in conventional fossil and nuclear fuel use.

Given that we are on a cultural trajectory still dominated by material temptations, significant energy inputs are needed to fuel the journey, especially as the nettle of population appears unlikely to be grasped any time soon, and our numbers are projected to increase by around 50 per cent by mid-century. Although the spectre of climate chaos stalks the periphery of our daily consciousness it has failed to trigger the kind of survival instincts we would see from a more tangible, physical and immediate danger. Consequently, little has so far been done at any great speed. This assumption of expanding energy needs calls for a new social foundation and a sudden social and political awakening, and massive commitment to decarbonize.

Our options then are centred on making our lifestyles, our way of making things and moving them, our way of doing business, less systemically dangerous through the widespread use of FITs. The solution is not, in the view of the authors, to leave the solutions to the conventional energy industry or even to other policy mechanisms such as quotas, credits and voluntary programmes. We believe decentralization and democratization of energy production to be a fundamental requirement for the 21st century, a shift in trajectory which will bring wide and deep benefits to those who participate. As the philosopher Jurgen Habermas once said, 'in the process of enlightenment there can be only participants'. The benefits of FITs are well proven, but they will only accrue significantly to the societies that choose to participate in this potential energy revolution.

These benefits can be quite substantial, for the advantages of renewable electricity democratization are economic, financial, environmental, social, political, geopolitical, technical and medical all at once, for they engender an energy system where:

- fuel is free, cheap, easy to find, and infinitely replenished once technology to exploit it is in place;
- supply is completely reliable and often indigenous, enhancing national security;
- the ever-present risk of resource conflicts is minimized as countries begin to use domestically available resources;
- almost any building, parking lot, roof, field or body of water can be used to generate electricity;
- the environmental burdens associated with electricity production, such as contaminated water, degraded land, spoiled air, a volatile climate, begin to recede; and
- economies become stronger by using more local employment, keeping revenues within the community and promoting a competitive manufacturing sector.

Indeed, these reasons imply that renewable energy *should* be much more advanced than it currently is. Lamentable political decisions, especially in the US and UK in the 1980s under the conservative Reagan and Thatcher governments – but still evident today – have picked conventional energy as the winner so far, and conventional resources continue to attract large subsidies from national governments. The United Nations Environment Programme (UNEP) report *Reforming Energy Subsidies: Opportunities to Contribute to the Climate Change Agenda* gives a figure of around US$300 billion (or 0.7 per cent of the world's economic output) going to such subsidies each year (United Nations Environment Programme, 2008).

A bias towards conventional energy sources exists not only at the national level, but the international one as well. The lack of a renewable energy counterweight to the International Atomic Energy Agency (IAEA), and the lack of consistent and sustained support for renewables from the International Energy Agency (IEA) has been a major obstacle to high-level political engagement with, and support for, such technologies. In fact, a group of experts from the Energy Watch Group produced a report accusing the IEA of publishing misleading data on renewables, and arguing that it has consistently underestimated the amount of electricity generated by wind power in its advice to governments (Peter and Lehmann, 2008). While it is debatable, they contend that the IEA shows 'ignorance and contempt' towards wind energy, while promoting oil, coal and nuclear as 'irreplaceable' technologies (Adam, 2009).

The establishment of the International Renewable Energy Agency (IRENA) in 2009 should help partially redress this, and ensure that renewable energy gets a better seat at the table when governments seek advice on energy strategy. Information provision, knowledge transfer, policy advice and capacity-building will all be part of IRENA's remit, drawing on the enormous wealth of existing expertise and experience around the world. This crucial new international body should make a rapid and effective contribution to renewables expansion, but will have to do so against continued opposition from the established interest groups that have limited the development of renewable energy until now. The unionized jobs and investments sunk in conventional energy generation creates momentum that influences government policies.

The good news is that this book is full of dozens of examples where leading nations in renewable energy, such as Germany and Spain, show that one can achieve more benefits from renewable energy resources than from conventional generation, and distribute those benefits more widely. This bundle of strengths is vital at this critical juncture when policy makers are once again debating the future of not only electricity supply but also of transport. Electricity generation may open the door to fundamental changes in our modes of transportation. The move to electric vehicles, for example, could substantially replace reliance on oil and gasoline with that on electricity (UK Industry Taskforce on Peak Oil and Energy Security, 2008). In addition, many technologies are coming forward which can generate electricity from vehicle-based infrastructure, such as magnetic levitation ('maglev') wind turbines near roadsides, heat and vibrations from road and parking surfaces, and magnetic generation from vehicles themselves using subsurface technology (Magkinetics, 2009).

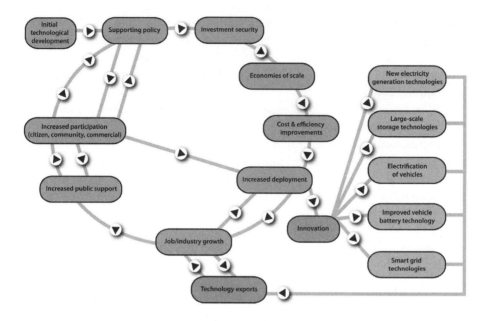

Figure 0.2 *The Industrial Revolution 2.0*

Source: Miguel Mendonça

There is no doubt that human ingenuity will lead to further breakthroughs of this kind, the speed of innovation corresponding with an increase in renewable energy deployment. Figure 0.2 gives an idea of the possibility of renewable energy technologies interacting with other energy systems to dramatically change the way we each conceive of, and use, energy. The examples on the right show the kinds of innovations developing at present, and they offer an indication of the scope of potential future development. So in a sense this book is about not only promoting the renewable energy technologies that exist today, but setting up a policy framework where they can be embraced tomorrow.

Collectively, the chapters found within this book will explore many of the remarkable benefits of well-designed FIT policies. These include cleaner and more efficient businesses and industries, empowering local citizens and politicians, promoting community cohesion and health, and reducing our dependence on fossil fuel combustion (See Table 0.1, and Chapters 2 and 3). The picture, however, is not entirely rosy, for poorly designed FITs bring with them a host of risks and challenges (See Table 0.2, and Chapter 4). The trick is, naturally, how to design FITs that accomplish the first set of benefits and minimize the second collection of costs. The following chapters tell you how.

- Chapter 1, 'The Green Economy', explores one of the most politically and socially pertinent issues for renewable energy policy today: economic recovery.

Table 0.1 *The economic, political, social and environmental advantages of FITs*

Economic	Create green-collar jobs
	Create domestic manufacturing and export industry
	Drive economic development
	Create hedge against conventional fuel price volatility
	Enable businesses, urban or rural, to develop new revenue streams
	Help to establish supply chains for renewable technologies
	Provide investor security
	Create stable conditions for market growth
	Drive down production costs of green electricity
	Develop and expand export opportunities in the renewable energy sector
	Simple, transparent policy structure helps encourage new start-ups and innovators
Political	Increase the stakeholder base supporting renewable energy policies
	Demonstrate commitment to renewable energy deployment
	Create mechanism for achieving renewable energy and emissions-reduction targets
	Increase understanding of potential citizen, community and business roles in environmental protection
	Increase energy security and energy independence
	Promote a more decentralized and resilient electricity system
Social	Encourage citizen and community engagement in activities protecting climate and environment
	Empower citizens and communities
	Increase resilience of communities
	Make renewable energy a common part of the landscape and cityscape
	Increased public support for renewables through direct stake and increased exposure to renewables
Environmental	Reduce carbon emissions
	Reduce pollution
	Encourage energy efficiency measures
	Reduce dependence on fossil fuels

It discusses the green economy, focusing on policy drivers, blockages to change, and why citizens and communities must be empowered to start producing their own energy.

- Chapter 2, 'Basic FIT Design Options', gives anyone that has never heard about FITs everything they need to know. It covers the basic definition of a FIT, how it works, and the basic elements of good FITs. This chapter enables policy makers around the world to draft a basic, well-functioning FIT scheme.
- Chapter 3, 'Advanced FIT Design Options', discusses all design options that may emerge after a FIT scheme has been in place for a number of years. A range of advanced design options can be implemented to further reduce windfall profits and to better integrate renewable electricity into the conventional power market.
- Chapter 4, 'Bad FIT Design', tells readers what design options should be avoided. Policy makers around the world have elaborated on FIT design for

Table 0.2 *The economic, political, social, and environmental challenges facing FITs*

Economic	Controlling costs of law if take-up is significant
	Lack of support from donors for FITs in developing countries
	Preventing 'gaming'
	Balancing investor confidence with cost control
	Price rises in raw materials for manufacturing technologies
	Securing supply chains for manufacturing
	Near-term upward pressure on electricity prices (if costlier resources such as solar PV are included in large amounts)
	Setting the prices accurately, and keeping them cost-efficient over time
	Tracking technological change can be challenging, particularly in emerging technologies
	Minimizing any negative economic effects of marginally higher electricity prices (e.g. sheltering electricity-intensive industries and low-income residents from full impact of FIT cost impacts)
Political	Opposition from vested interests
	Low government prioritization of renewable energy
	Prioritization of large centralized supply, such as nuclear and 'clean' coal
	Pressure to design 'low-impact' (ineffective) FITs
	Combining with existing schemes
	Transition from existing schemes without investment disruption
	Ensuring planning system does not hamper development
	Opposition to higher electricity prices
	Permitting and other barriers
Social	Competition for best sites can lead to 'NIMBYism'
	Opposition to cost rises in fuel bills (especially for those in fuel poverty)
	Creating investment vehicles for average citizens to participate (e.g. low interest loans, etc)
	Maximizing local participation and ownership, rather than solely corporate and utility ownership
	Sheltering low-income residents from near-term impacts on electricity prices
	Cost sharing across customer classes, and geographic areas
	New transmission lines can generate opposition to projects
Environmental	Limitations of available sites for installations
	Air quality concerns in the case of biomass combustion systems
	Marine impacts in the case of tidal, wave and offshore wind
	Species-specific concerns (bats, birds, etc)
	Land-use issues and conflicts
	New transmission lines can create opposition
	Sound and 'strobe' impacts (specifically for wind turbines)

more than three decades, not always with success. In order to avoid such pitfalls, we have used several case studies to describe what has turned out to be worst practice.

- Chapter 5, 'FIT Design Options for Emerging Economies', includes special, innovative design options for emerging economies and developing countries. FIT design is very flexible in that it can accommodate the unique needs for developing countries alongside developed ones. This chapter places an emphasis on issues relating to funding, cost control and mini-grids.

- Chapter 6, 'FIT Developments in Selected Countries', explores the actual use of FIT programmes in a representative sample of countries and regions. The chapter includes case studies from Europe (Germany, Spain, and the UK), North America (Canada and the US), Africa (Kenya and South America), Asia (India) and Australia.
- Chapter 7, 'Dispelling the Myths about Technical Issues', destroys a number of pernicious myths surrounding renewable energy. It documents how many renewable energy systems already provide reliable, around-the-clock electricity. It shows that 'variable' renewable resources such as wind and solar can become highly reliable when interconnected with each other, with other renewables, or integrated with energy efficiency and storage technologies. It argues that the so-called 'technical issues' relating to intermittency, interconnection, and transmission and distribution no longer need prevent rapid renewable expansion.
- Chapter 8, 'Barriers to Renewable Energy Development', takes a closer look at the economic, political and social impediments to widespread use of renewable electricity. These include lack of information and market failures, inconsistent political support and obscene subsidization of conventional energy systems, and cultural attitudes and values prioritizing consumption and abundance over efficiency and frugality.
- Chapter 9, 'Other Support Schemes', takes a hard look at eight types of policy mechanisms often used in conjunction with (and sometimes instead of) FITs: renewable portfolio standards and quotas, tradable certificates and Guarantees of Origin, voluntary green power programmes, net metering, public research and development expenditures, system benefits charges, tax credits, and tendering. The chapter concludes by explaining how FITs have advantages over each of them.
- Chapter 10, 'Campaigning for FITs', provides some guidelines and pointers on how FITs can be successfully pushed and politically accepted. It also looks at what stands in their way. The chapter should be considered in conjunction with Chapter 8 on barriers. Experience has shown that with the right mix of tactics, arguments, and human and financial resources, the overwhelming logic in favour of using FITs can trump any opposition to them, despite challenges along the way.

Let us pre-empt a common objection to FITs here. When looking at Tables 0.1 and 0.2, some astute readers may ask how it is that well-designed FITs actually reduce electricity prices if they pay renewable producers more for their electricity. This is a good question, but it has a simple answer.

From a macroeconomic and long-term perspective the deployment of renewable energies brings about many benefits, which far surpass the initial short-term costs related to tariff payment. In some countries, such as Germany and Spain, this can happen (benefits outweighing costs) quite quickly. The Spanish case has shown for the past few years that the massive deployment of cheap wind energy reduces the overall costs of electricity. By 2007, wind turbines and wind farms in Spain produced 26.7TWh of electricity which cost consumers about €1 billion. At the same time, the large share of wind power traded at the electricity spot market, brought about by FITs, reduced the market price by €0.006/kWh, saving utilities and consumers about €1.7 billion in avoided costs (for a net savings of more than €640 million) (Gasco, 2008). This so-called merit-order effect (see Section 6.1.1) was also observed in Germany, another country with a large share of renewable electricity in the national electricity portfolio. The German Ministry for the Environment has calculated that their FIT cost electricity customers €3.2 billion in 2007, but saved them more than €5 billion through the merit-order effect.

The icing on the cake is that these calculations did not even take the numerous additional benefits of renewable energies into account. FITs all over the world have stimulated the job market and led to important reductions in greenhouse gas emissions. An important example is given by Germany where their FIT created more than 280,000 jobs and cut greenhouse gas emissions by 79 million tonnes of carbon dioxide equivalent (CO_2e) (for a detailed cost–benefit analysis of the German FIT scheme, see Section 6.1.1). In addition, it has put the Germans well ahead of their renewables deployment targets.

Taken together, these chapters will clearly show that a great change in the world's energy sources is needed, and this book examines the most proven tool for getting the job done, the Swiss army knife of renewable energy policies – the FIT. The purpose of the authors in putting the book together is not to be uncritical cheerleaders; we have no shares in renewable energy companies or financial interests connected with FITs of any type. What we have concluded is that FITs allow the kind of collaboration that will necessarily become fundamental to the process of solving climate, environment and social justice issues this century. They bring so many benefits that they are, in essence (and much like investments in energy efficiency), a free lunch you get paid to eat.

For these reasons, and all those that faithful readers are about to discover, we commend FITs to policy makers, governments, campaigners, investors, businesses and anyone who cares about the future and wants to take action today to protect it.

References

Adam, D. (2009) 'International Energy Agency "blocking global switch to renewables"', *The Guardian*, 9 January, www.guardian.co.uk/environment/2009/jan/08/windpower-energy

European Renewable Energy Council (2007) *Future Investment: A Sustainable Investment Plan for the Power Sector to Save the Climate*, European Renewable Energy Council, Brussels

Gasco, C. (2008) *Economic Impact of Renewable Energy Expansion*, presentation at the fifth workshop of the International Feed-in Cooperation, Brussels

Magkinetics (2009) 'Technology', www.magkinetics.com/technology.html

Peter, S. and Lehmann, H. (2008) 'Renewable energy outlook 2030 – Energy Watch Group Global renewable energy scenarios', Energy Watch Group, www.energywatchgroup.org/fileadmin/global/pdf/2008-11-07_EWG_REO_2030_E.pdf

UK Industry Taskforce on Peak Oil and Energy Security (2008) 'The oil crunch – Securing the UK's energy future', http://peakoiltaskforce.net/wp-content/uploads/2008/10/oil-report-final.pdf

United Nations Environment Programme (2008) 'Reforming energy subsidies: Opportunities to contribute to the climate change agenda', www.unep.org/pdf/PressReleases/Reforming_Energy_Subsidies.pdf

1

The Green Economy

'The green economy' is a somewhat bemusing term from the perspective of the ecologist, the holistic thinker. No economic activity can take place without drawing on raw materials provided by nature, and neither can any kind of human or animal existence. That the environment has for so long been conceptually and legislatively relegated to the periphery of the economy is a major cause of the variety of 'crunches' we find ourselves faced with – financial, economic and environmental. Our very lifestyles are predicated culturally and economically upon ever-increasing resource use, treating constant material acquisition (and the inefficiency, waste, and disposal it involves) as a norm.

First and foremost, this is leading *us* – the adults of today – into trouble, and leaving us with thinning options. Second, if we truly mean what we say about our love and concern for our children, we need to concentrate harder on what our actions – and our values – mean for them, and their children, ad infinitum. Our individualism and materialism exhibit perilous consequences, for both the present and the future.

The world has been shaped by the decisions made not only by the powerful, but by all of us, aggregating over the centuries. However ethical we may like to think ourselves, we are inextricably bound up in a system which creates, by its logic and values, increasingly dangerous environmental and social problems. However unbelievable it may seem, all credible evidence shows that our uncritically materialist way of life threatens the very existence of human society, and life on Earth. Hyper-consumption, ignorance of the natural world, and a disregard for ecological limits have become naturalized into our lifestyles. The ability to resist the dictates of the most powerful interests, and values predisposed against consumption and environmental destruction, are challenges to our ingenuity, determination, and strategic and critical thinking. But they also require cooperation, something which humans can excel at.

Changing the world is therefore both necessary and possible. This is the good news, and it is not lost on a great many people. At the time of writing, a 'historic' meeting of the Group of Twenty (G-20) heads of government has taken place in London. They have strived for unity, from many different entry points – often contradictory ones. Outside the proceedings, a coalition of many disparate groups called 'Put People First' protested. Their causes are many, but their message was 'Jobs, Justice, Climate'. The interconnection of these issues centres on values, and is explicit in their stated aim of putting people and the natural world before the economy. Both the politicians and the protesters are in effect working towards the same things, although they may not realize it, as the *realpolitik* within the conference shapes the truth in different ways.

But where would they explicitly agree? In this time of crises, both top-down and bottom-up representatives are surely going to find consensus on certain responses. One of the clearest examples is from America. Although far from alone, three African American men in particular have championed a vision and a message: greening the economy can protect people and jobs, not harm them. These men are Van Jones, founder of Green For All; Jerome Ringo, President of The Apollo Alliance; and Barack Obama, President of the US. They have argued that the opportunities presented by these crises can be solved in large part by the same thing – decarbonizing the economy. Not only will this create new technologies, industries and jobs, it can also provide a clean, safe environment and improved domestic energy security. There is not as much mention of it preventing the need for further military adventures in the Middle East for example, but that cannot be lost on such proponents – particularly when those costs are estimated to be several billion dollars per month, costing up to US$2.7 trillion by 2017 (Associated Press, 2008). President Bush refused to sign the Kyoto Protocol, suggesting it would cost the economy nearly $400 billion and five million jobs (Heilprin, 2004). It may well be a mere coincidence that as Barack Obama campaigned for the Presidency, he and Joe Biden promised to create five million *new* jobs through green measures.

Through what some might consider a kind of divine intervention, the US finally got a leader with the right ideas at the right time. President Obama said in an address:

> *We'll put people back to work rebuilding our crumbling roads and bridges, modernizing schools that are failing our children, and building wind farms and solar panels, fuel-efficient cars and the alternative energy technologies that can free us from our dependence on foreign oil and keep our economy competitive in the years ahead.* (Obama, 2008)

And indeed he has put American money where his mouth is, and pledged huge sums for green measures. After a battle with Republican opposition, the American Recovery and Reinvestment Act 2009 was passed. Table 1.1 shows the green part of the stimulus package in full, totalling almost $113 billion over two years.

This package also provided the funds that had been authorized by the Green Jobs Act of 2007 (H.R. 2847), which called for $125 million in funding to establish national and state job training programmes to help address skills shortages that

Table 1.1 *Clean energy details of American Recovery and Reinvestment Act*

$4 billion	For job training, with focus on green-collar jobs
$32 billion	To transform the nation's energy transmission, distribution and production systems by allowing for a smarter and better grid and focusing investment in renewable technology
$11 billion	Reliable, efficient electricity grid
$6 billion	To weatherproof modest-income homes
$31 billion	To modernize federal and other public infrastructure with investments that lead to long-term energy cost savings
$20 billion	For local school districts through new School Modernization and Repair Program to increase energy efficiency
$16 billion	To repair public housing and make key energy efficiency retrofits
$1 billion	Public Housing Capital Fund for projects that improve energy efficiency
$1.5 billion	HOME Investment Partnerships to help local communities build and rehabilitate low-income housing using green technologies
$6 billion	GSA Federal Buildings for renovations and repairs to federal buildings to increase energy efficiency and conservation
$6.9 billion	Local Government Energy Efficiency Block Grants to help state and local governments become energy efficient and reduce carbon emissions
$2.5 billion	Energy Efficiency Housing Retrofits for a new programme to upgrade HUD-sponsored low-income housing to increase energy efficiency
$2 billion	Energy Efficiency and Renewable Energy Research for development, demonstration, and deployment activities to foster energy independence, reduce carbon emissions and cut utility bills
$500 million	For advanced energy efficient manufacturing
$1.5 billion	Energy Efficiency Grants and Loans for Institutions for energy sustainability and efficiency grants to school districts, institutes of higher education, local governments and municipal utilities
$500 million	Industrial Energy Efficiency for energy efficient manufacturing demonstration projects
$10 billion	For transit and rail to reduce traffic congestion and gas consumption
$2 billion	Advanced Battery Loans and Grants to support US manufacturers of advanced vehicle batteries and battery systems
$1.1 billion	Amtrak and Intercity Passenger Rail Construction Grants
$200 million	Electric Transportation for a new grant programme to encourage electric vehicle technologies
$2 billion	To support advanced battery development
$2.4 billion	For carbon sequestration research and demonstration projects
$1.85 billion	For various clean energy projects to promote energy smart appliances; assist states and GSA to convert fleets to more efficient vehicles; electric vehicle technology research; developing renewable energy for military use
$400 million	Alternative Buses and Trucks to state and local governments to purchase efficient alternative fuel vehicles
$8 billion	Renewable Energy Loan Guarantees for alternative energy power generation and transmission projects
$350 million	Department of Defense research into using renewable energy to power weapons systems and military bases
$2.4 billion	Cleaning Fossil Energy for carbon capture and sequestration technology demonstration projects
$400 million	For NASA climate change research

Source: Schneider, 2009

Table 1.2 *Economic stimulus packages and green portions*

Country/region	Stimulus (US$ billion)	Green portion	Green percentage
Australia	26.7	2.5	9
Canada	31.8	2.6	8
China	586.1	221.3	38
EU	38.8	22.8	59
France	33.7	7.1	21
Germany	104.8	13.8	13
Italy	103.5	1.3	1
Japan	485.9	12.4	3
South Korea	38.1	30.7	81
UK	30.4	2.1	7
US	972.0	112.3	12

Source: Adapted from *Financial Times*, 2009

threaten to slow growth in green industries. This capacity-building is essential for a green economy.

The US was not the only nation to take the opportunity to raise funds for green measures in this time of chaos. The anglophone world, European nations, the European Union (EU) as a bloc, and major Asian industrial nations all became involved (see Table 1.2).

National governments, UNEP, international trade union federations, think tanks, non-governmental organizations (NGOs) and others have also attempted to set out the overwhelming logic of meeting our greatest threats with the same solution.

Something even more amazing about these multi-billion dollar aid packages: they barely scraped the surface in terms of capturing the true potential of renewable electricity resources. The US Department of Energy, hardly an organization biased in favour of renewable energy, calculated that the amount of wind and solar resources found within the US amounted to a resource base the equivalent of more than 300,000 billion barrels of oil, or more than 20,000 times the annual rate of national energy consumption at that time (US Department of Energy, 1989). In early 2009, a peer-reviewed academic study estimated that the world has roughly 3,439,685TWh of untapped solar, wind, geothermal, and hydroelectric potential – about 201 times the amount of total electricity consumed in 2007 (Jacobson, 2009). Much of this potential, furthermore, is economically feasible. One global assessment found that the world could *double* its expected investments in renewable electricity supply from 2004 to 2030, an amount greater than $5.6 trillion of new (and profitable) investments (See Figure 1.1) (European Renewable Energy Council, 2007). In the US, wind and solar electricity systems could have met three times the total amount of installed capacity operating in 2008 with existing, commercially available technology (Sovacool and Watts, 2009).

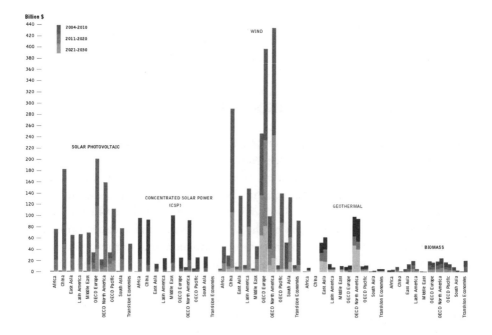

Figure 1.1 *Expected investments (left column) versus investment potential (right column) for renewable electricity systems around the world, 2004–2030*

Note: expected investments are those that are expected to occur under a 'business as usual' case from 2004 to 2030; investment potential is the full potential that could be achieved if regulators aggressively pursued renewables and invested in every single opportunity that was cost-effective.

Source: European Renewable Energy Council, 2007

However, the billions of dollars being poured into clean energy, the trillions of dollars of potential investment opportunity, better transport infrastructure, capacity-building and so on has less to do with environmental wisdom than with financial and political well-being and sustainability. It may have even *more* to do with the enormous potential of export markets for green materials and technologies. One could trace some of the German enthusiasm for renewable energy support to the vivid pictures that must have formed in the minds of legislators as they considered the market capture potential they could unleash. Their nuclear phase-out plans, along with their R&D and manufacturing capacity, opened the door to incredible possibilities. These have been realized with some precision, despite some epic battles with entrenched interests. As described above and especially in Chapter 10, it takes some doing to overcome the existing order, but coalitions promoting the interests of society can prevail, and we now hold this country up as probably the world's greatest example of renewable energy policy and technology.

It is not just in renewables that they have excelled, but in greening their economy in many areas. When we think of 'green-collar jobs' we are not thinking only of solar panel and wind turbine designers and installers. Green jobs are vastly deeper and wider than that. The definition is hard to nail down, and is composed of deceptively simple ideas.

1.1 Defining Green Jobs

'Green collar' as a term first appeared at a 1976 Congressional hearing in the US, when Professor Patrick Heffernan delivered a paper entitled *Jobs for the Environment – The Coming Green Collar Revolution*. Not much had been heard of the concept until relatively recently, and mostly in the US again. Various individuals and groups including Van Jones, Green For All; the Apollo Alliance; the Blue Green Alliance; and Obama, Clinton and McCain on the 2008 campaign trail have all popularized the term. They have linked the health of the economy to that of the environment, and the causes of poverty to environmental pollution. They have dispelled the myth that these issues are separate.

Ultimately, *all* jobs should be 'green' jobs, but at present, the term prompts the question – how green can a job be in a dirty economy? In a major UNEP report on the issue, the authors refer to 'shades of green':

> *A green economy is an economy that values nature and people and creates decent, well-paying jobs. Technological and systemic choices offer varying degrees of environmental benefit and different types of employment. Pollution prevention has different implications than pollution control, as does climate mitigation compared with adaptation, efficient buildings vis-à-vis retrofits; or public transit versus fuel-efficient automobiles. It is of course preferable that the most efficient, least polluting options receive priority. But these are not either-or choices, as all of these options are needed to bring about a more sustainable, low-carbon economy. But they do suggest 'shades of green' in employment. Greater efficiency in the use of energy, water, and materials is a core objective. The critical question is where to draw the line between efficient and inefficient practices. A low threshold will define a greater number of jobs as green, but may yield an illusion of progress. In light of the need to dramatically reduce humanity's environmental footprint, the bar needs to be set high – best available technology and best practices internationally should be seen as the most appropriate thresholds. And, given technological progress and the urgent need for improvement, the dividing line between efficient and inefficient must rise over time. Hence, 'green jobs' is a relative and highly dynamic concept* (UNEP, ILO and ITUC, 2007, ppxi and xiii)

Green-collar jobs can exist in many different sectors, across all levels of skill and responsibility. They may be in notably green fields such as renewables or energy efficiency, or be 'greener' versions of traditional sectors – architecture, construction, manufacturing or finance. In his recent book, Van Jones defines a green-collar job as 'a family-supporting career-track job that directly contributes to preserving or enhancing environmental quality' (Jones, 2008). He sets out three core principles which define a green economy: equal protection for all; equal opportunity for all; reverence for all creation (Jones, 2008). The equal opportunities arguments are common in academia and advocacy in the US. Attention is drawn to the fact that low-income workers, and those who have difficulties finding work, can gain attractive, meaningful employment and career prospects in the expanding green-collar sector due to the low barriers to entry (Pinderhughes, 2007). So, in the absence of a universally agreed definition, there are several common features. Green-collar jobs:

- are related to environmentally friendly products and services;
- are relevant to all education and skill levels;
- provide a living wage and health benefits;
- offer career development; and
- are often locally based.

1.2 Policy Drivers

Renewable energy has so far been one of the highest job-creating sectors. An impressive list of facts and figures from the US and Germany shows the kind of economic and employment opportunities that could be and are being seized in these areas. According to the American Solar Energy Society, the US has a huge industry in renewables and efficiency already, and this could grow enormously in the coming years with the right incentives:

> *We found that, in 2007, the US RE&EE [renewable energy and energy efficiency] industries generated $1,045 billion in sales and created over 9 million jobs – including $10.3 billion in sales and over 91,000 jobs in Colorado. The US RE&EE revenues represent substantially more than the combined 2007 sales of the three largest U.S. corporations – Wal-Mart, ExxonMobil, and GM ($905 billion). RE&EE are growing faster than the US average and contain some of the most rapidly growing industries in the world, such as wind, photovoltaics, fuel cells, recycling/remanufacturing, and biofuels. With appropriate federal and state government policies, RE&EE could by 2030 generate over 37 million jobs per year in the US.* (American Solar Energy Society and Management Information Services, Inc., 2008)

Another report forecast that, with the appropriate public policies in place, by 2030 as many as one out of four workers (40 million people) in the US will be working in the areas of renewables and efficiency, industries which will be worth up to $4.5 trillion in revenue. They claim that these industries already turn over nearly $1 trillion, generating more than $150 billion in tax revenues (American Solar Energy Society, 2007).

Yet another study on job creation from renewables found that the renewables industry provides more jobs per unit of delivered energy than the fossil fuel industry. Further, it suggested that it is the comprehensiveness and coordination of energy policy that yields the biggest combined rewards for the various sectors (Kammen et al, 2004).

Policy drivers have been most apparent in renewables, but are now spreading into many other key areas, including energy efficiency, conventional energy generation, manufacturing, waste, buildings and transport:

- extended producer responsibility (product take-back, reuse and recycling laws);
- public and private sector green procurement (mandates for purchasing eco-friendly products and services);
- eco-labelling (guiding purchasing choices by creating standards);
- recycling and anti-landfill mandates (e.g. obligations upon local authorities);
- green building standards (e.g. UK's Zero Carbon Homes by 2016 policy);
- energy efficiency retrofits (government-funded schemes, mandates);
- sustainable transport (walking and cycling promotion, alternative fuel mandates, tram or bus rapid transit systems); and
- renewable energy and energy efficiency targets, mandates and incentives (feed-in tariffs (FITs), solar roof programmes, solar thermal ordinances, tax credits, portfolio standards, fuel efficiency standards).

These policy drivers are already proving successful, with global markets for renewables estimated at around €1000 billion, and projections showing that revenues of €2200 billion are likely to be realized in the next decade:

- energy efficiency (appliances, industrial processes, electrical motors, insulation, etc): €450 billion at present (€900 billion by 2020);
- waste management/recycling: €30 billion (€46 billion by 2020);
- water supply/sanitation/water efficiency: €185 billion (€480 billion by 2020);
- sustainable transport (more efficient engines, hybrids, fuel cells, alternative fuels, etc.): €180 billion (€360 billion by 2020) (UNEP, ILO and ITUC, 2007, p16).

As Section 6.1.1 of this book shows, Germany has become an icon of renewable energy job creation for the last decade. Elsewhere, the US, China, Spain and India are leading the field globally, with France, Portugal, Brazil and Japan also achieving great progress. As Table 1.3 shows, there is not much correlation between country size, gross domestic product (GDP) and renewable energy success. It is more dependent on the combination of renewable energy resources on one hand, and political determination and social awareness on the other. The determination to exploit these resources will ensure that prioritization is given financially and supported in policy.

Targets and policies for renewables deployment send vital signals to the market, and all those who wish to participate in it. Over 70 countries worldwide now have targets for deployment. The US and Canada do not have national targets, but many US states and Canadian provinces have established their own. The EU has country-specific targets to be reached by 2020. But targets can be toothless without the policies and supporting conditions to reach them, and penalties for not doing so.

The majority of countries seeking a policy to rapidly increase their renewable energy capacity employ the FIT model. It has driven the deployment of more renewable capacity, and at lower cost, than any other. The academic literature is quite unanimous in its appraisal of the efficiency and effectiveness of the various support schemes, as we explore in Chapter 9. *The Stern Review: The Economics of Climate Change* confirmed this, through reviewing the literature comparing support schemes (Stern, 2006). Personal research has found the same (Mendonça, 2007). Stern reiterates the German case as a policy success story in his new book on addressing climate change, *A Blueprint for a Safer Planet* (Stern, 2009).

Table 1.3 *Top five renewable energy nations: selected indicators*

Country	Renewable energy production (GW)						Area (sq miles)	GDP (per capita)
	Wind	PV	Biomass	Small hydro	Geo	Total		
China	12.2	>0.1	3.6	65.0	~0	81	9,596,960	$6000
USA	24.2	0.8	8.0	3.0	3.0	40	9,826,630	$47,000
Germany	23.9	5.4	3.0	1.7	0	34	357,021	$34,800
Spain	16.8	2.3	0.5	1.8	0	21	504,782	$34,600
India	9.7	~0	1.5	2.0	0	13	3,287,590	$2800

Notes: Total includes other technologies such as CSP and solar thermal.
Source: REN21, 2009; Central Intelligence Agency, 2008

1.3 BLOCKERS

Until recently, many anglophone countries have favoured so-called market-based mechanisms and rejected FITs. They did so for a wide variety of reasons, but mainly because they do not support the practice of 'price fixing'. It has been said that they deem FITs to be a socialist idea. However, the overwhelming evidence in support of the efficiency and effectiveness of FITs has helped many pro-renewables coalitions to push these ideological barriers aside. How should setting the price be more or less market intervention than setting a quantity? It helps that there exists the 'premium' feed-in model, which is considered more market-oriented than the straight German-style 'tariff' model (see Section 3.1).

But in all likelihood, this professed ideology is unlikely to be the real reason behind opposition to FITs. The alternative mechanisms favour large, credit-worthy investors and utilities – who are often the monopoly suppliers of energy. Indeed, other support schemes tend to limit the growth of renewable energies, thus guaranteeing large market shares for conventional energy sources which are generally in the hands of large private corporations or oligopolists. We go back to the theme of power and influence. These groups have the ability to guide national public policy in their favour, which can result in a policy landscape that offers little to those wishing to enter the market, or even find a cost-effective solution for putting solar panels on their house or business. 'Green certificate' trading schemes are a classic example. They tend to pay high prices to large developers, while taking no account of the needs of smaller players. Further, they can provide a disincentive to deployment, as the fewer certificates on the market, the higher their individual value (something we explore in detail in Chapter 9 on other support mechanisms).

Tradable certificates, along with other dodgy and defunct mechanisms, are promoted and defended by large energy companies and monopolies, and the same is true in the US with Solar Renewable Energy Certificates (SRECs). Ironically, opponents of FITs tend to accuse them of having the failings of the schemes they support, such as high cost, complexity and the prevention of innovation. This is the perfect time to uncover the truth and think about how one might analyse policies to see who really benefits from them. Given what is at stake, we must avoid the potential for self-interested organizations to prevent citizen and community participation in the energy market. The importance of this cannot be overstated. These themes are explored further in Chapter 10.

1.4 Why Citizens and Communities Must Join In

Danish and American experience shows that renewable energy deployment is accelerated or held back depending on whether policies allow or prevent investment and participation on the part of the general public. Public engagement and acceptance, for example, have been shown to be clearly linked to the nature of investment possibilities (Mendonça et al, 2009). From the 1970s, Danish community wind partnerships became increasingly common, with local people pooling their financial resources to invest in their own wind farm. When this model broke down due to the change to less favourable policies, and was replaced by larger, purely business investments, opposition to wind power development increased, as the local population no longer had a stake in the wind energy business (Girardet and Mendonça, 2009).

There are two primary reasons for facilitating citizens and communities in engagement with climate- and environment-protecting activities. The first is about winning the battle for hearts and minds. Even with a massive uptake of energy efficiency measures throughout our economies, we still face a growing global population – perhaps nine billion by 2050, with growing needs and wants (and we have not created an ethical, sustainable world with three or six billion people). Much of our economic growth of the last two decades has occurred in the context of a boom in telecommunications and consumer electronics. The success of these industries has become evident in the 'dependency creep,' which has emerged. Our lives have become increasingly dependent upon electronic communications systems, global positioning systems (GPS) and other mobile devices. There has even occurred the recognition of an anxiety condition called 'nomophobia', the fear of being without a mobile phone signal. MP3 music players have become incredibly popular, selling hundreds of millions of units annually around the world. The same is true of video game consoles of all types, the industry being reputedly worth more than the film industry. These systems enable people to distance themselves from the natural environment, and make it more difficult to reflect on the energy implications of building lifestyles dependent on energy-intensive technology. For the sake of irony, it should be noted that this book was produced by three authors in different parts of the world who never met during the process, and could research and share their thoughts at high speed via the internet and telephone.

So, we have high technology and its energy needs on the one hand, and an 'environmentocidal' energy system on the other. If we had another hand to point the way to a third path, it would be illustrating the need for renewable energy. Given the massive social and environmental costs associated with nuclear power (Sovacool and Cooper, 2008), renewables are pretty much all that we have to address society's daunting energy challenges in the near term. Ed Miliband, the British Energy and Climate Change Secretary, has said that opposing wind farms should be socially taboo – as unacceptable as not wearing a seatbelt (Stratton, 2009).

Opposition to wind farms is generally about views towards landscapes, and not generally about views against renewable energy. The 2009 climate change docu-drama *The Age of Stupid* has a thread featuring proponents and opponents of a wind farm. Protesters comment on concerns about noise from the turbines as we cut to footage from Santa Pod Raceway, a nearby racetrack which is evidently one of the noisiest places in Britain. A memorable interview with a victorious anti-wind farm protester shows her experiencing an on-camera logic meltdown as she celebrates the victory over wind energy while voicing support for renewable energy in principle.

So the first point is that public support is vital for the smooth transition to a renewables-based energy system. The second point concerns what else can happen when the public, as well as businesses, factories, local authorities, farms, hospitals, schools and so on are able to get hands-on experience with renewable energy. Quite simply, they then have a stake in the green economy, and further legislative greening – depending on the specifics – is more likely to be welcomed. You can see a disempowered, disinterested populace become active, aware partners in remaking some of the fundamental aspects of society. Feed-in tariffs offer such public engagement in a simple way.

Importantly, feed-in tariffs can also help bring disparate renewable energy industries together, instead of leaving them to scrap for support from a variety of grants and other schemes. There is nothing to be gained from criticizing one's natural allies, but it does happen: partly, one might conclude, due to the lack of a universally accepted simple, inclusive, fair scheme like a FIT. Getting organized and speaking with one voice is essential for this industry, if only to offset the powerful and organized lobbies behind conventional energy systems such as coal and nuclear.

To conclude, let us remind readers of a few central points. A green economy, one based on the principles of ecology and appreciative of limits, is not at odds with better jobs, macroeconomic stability, or improved standards of living. Indeed, these factors are seamlessly joined: the clean energy promoted by FITs simultaneously strengthens local employment, stabilizes energy prices, and promotes energy literacy, democracy and (at times) activism. FITs are not the only way to achieve a green economy, but they are certainly one of the most direct and cost-effective mechanisms to start us on that pathway. Embracing them is an essential first step towards creating an economy that preserves the planet for future generations instead of ruining it to satisfy today's desires.

REFERENCES

American Solar Energy Society and Management Information Services, Inc. (2008) *Defining, Estimating, and Forecasting the Renewable Energy and Energy Efficiency Industries in the U.S. and in Colorado*, New York, NY

American Solar Energy Society (2007) *Renewable Energy and Energy Efficiency: Economic Drivers for the 21st Century*, American Solar Energy Society, Boulder, CO

Associated Press (2008) 'Studies: Iraq war will cost $12 billion a month', www.msnbc.msn.com/id/23551693/, 9 March

Central Intelligence Agency (2008) *The World Factbook*, www.cia.gov/library/publications/the-world-factbook/

European Renewable Energy Council (2007) *Future Investment: A Sustainable Investment Plan for the Power Sector to Save the Climate*, European Renewable Energy Council, Brussels

Financial Times (2009) 'Which country has the greenest bail-out?', www.ft.com/cms/s/0/cc207678-0738-11de-9294-000077b07658.html?nclick_check=1, 2 March

Girardet, H. and Mendonça, M. (2009) *A Renewable World: Policies, Practices and Technologies*, Green Books, Totnes

Heilprin, J. (2004) 'Bush stands by rejection of Kyoto Treaty', www.commondreams.org/headlines04/1106-07.htm, Associated Press, 6 November

Jacobson, M. Z. (2009) 'Review of solutions to global warming, air pollution, and energy security', *Energy and Environmental Science*, vol 2, pp148–173

Jones, V. (2008) *The Green Collar Economy*, Harper One, New York, NY, p12

Kammen, D., Kapadia, K. and Fripp, M. (2004) *Putting Renewables to Work: How Many Jobs Can the Clean Energy Industry Generate?* Renewable and Appropriate Energy Laboratory, University of California, Berkeley, CA

Mendonça, M. (2007) *Feed-in Tariffs: Accelerating the Deployment of Renewable Energy*, Earthscan, London, pp17–18

Mendonça, M., Lacey, S. and Hvelplund, F. (2009) 'Stability, participation and transparency in renewable energy policy: lessons from Denmark and the United States', *Policy and Society*, vol 27, pp379–398

Obama, B. (2008) 'President-Elect Obama's weekly democratic radio address – 2.5 Million Jobs', my.barackobama.com/page/community/post/stateupdates/gGxtlN, 22 November

Pinderhughes, R. (2007) *Green Collar Jobs: An Analysis of the Capacity of Green Businesses to Provide High Quality Jobs for Men and Women with Barriers to Employment*, Executive Summary, pp3–4, http://blogs.calstate.edu/cpdc_sustainability/wp-content/uploads/2008/02/green-collar-jobs_exec-summary.pdf

REN21 (2009) *Renewables Global Status Report: 2009 Update*, REN21 Secretariat, Paris, www.ren21.net/globalstatusreport/g2009.asp

Schneider, K. (2009) 'Clean energy is foundation of proposed stimulus', apolloalliance.org/new-apollo-program/clean-energy-serves-as-foundation-for-proposed-reinvestment-bill/

Sovacool, B. K. and Cooper, C. (2008) 'Nuclear nonsense: Why nuclear power is no answer to climate change and the world's post-Kyoto energy challenges', *William and Mary Environmental Law and Policy Review*, vol 33, no 1, pp1–119

Sovacool, B. K. and Watts, C. (2009) 'Going completely renewable: Is it possible (let alone desirable)?', *The Electricity Journal*, vol 22, no 4, pp95–111

Stern, N. (2006) *Stern Review: The Economics of Climate Change*, Cambridge University Press, Cambridge, p366

Stern, N. (2009) *Blueprint for a Safer Planet*, The Bodley Head, London, p116

Stratton, A. (2009) 'Opposing wind farms should be socially taboo, says minister', *The Guardian*, 24 March

UNEP, ILO and ITUC (2007) *Green Jobs: Towards Sustainable Work in a Low-Carbon World*, United Nations Environment Programme, International Labour Organization and International Trade Union Confederation, 21 December

US Department of Energy (1989) *Characterization of US Energy Resources and Reserves*, DOE/CE-0279, Washington, DC

2
Basic FIT Design Options

This chapter presents the most commonly used design options for feed-in tariffs (FITs). With these in mind, any eager person will be able to draft a basic but complete FIT scheme for their country or region. Where appropriate, we have included examples from regions or countries with particularly good designs. More complex and less frequent design options are presented in the next chapter on advanced FIT design options. These might be of special interest for those living in countries that have already operated a FIT scheme for several years. In addition, unsuccessful design options are presented in chapter 4 on bad FIT design.

When designing FITs, the idea is to provide a balance between investment security for producers on the one hand and the elimination of windfall profits (in order to reduce the additional costs for the final consumer) on the other. To accomplish these goals, the design of FITs has therefore become increasingly complex. Empirical studies, however, have identified a number of design criteria which legislators and the rest of us should consider when drafting or improving FIT legislation (Mendonça, 2007; Roderick et al, 2007; Sösemann, 2007; Grace et al, 2008; Klein et al, 2008; Fell, 2009b). Chapters 2–4 are based on ongoing research in the framework of the PhD project of David Jacobs, sponsored by the German Federal Foundation for the Environment (DBU). The checklist at the end of this chapter will help readers who may get lost in the details.

In countries with a relatively short history of renewable energy development, and those that are establishing a FIT scheme for the very first time, we recommend keeping the support mechanism simple at the start. It should be easy to understand as FITs 'invite' all parts of a society to become electricity producers, ranging from private households to large utilities. Therefore, the legislation should be comprehensible to anyone without the assistance of legal experts. At a later stage, the FIT might have to become more complex, but by then producers will have become experienced with this type of support scheme. A good example of this increase in complexity over time is the German FIT scheme. While the first FIT law from 1990 included only 5 articles, the number increased to 13 in 2000, 21 in 2004, and a staggering 66 articles in 2009. This was to account for issues connected to better market integration, grid connection, and tariff differentiation (among others).

As this chapter will show, basic FIT design options include:

- eligible technologies;
- eligible plants;
- tariff calculation methodology;
- technology-specific tariffs;
- size-specific tariffs;
- duration of tariff payment;
- financing mechanisms;
- purchase obligations;
- priority grid access;
- cost-sharing methodology for grid connection;
- effective administrative procedures;
- setting targets; and
- progress reports.

Let us begin with the most basic element of any FIT programme, eligible technologies.

2.1 Eligible Technologies

As a first step, legislators will have to decide which renewable energy technologies they want to support, i.e. which technologies will be eligible for tariff payment under the FIT scheme. In order to make this decision there should be good knowledge about the potential and resource availability of each technology in a given region or country. National wind and solar maps (along with other resource maps) can be very useful for this purpose.

Generally, it is recommended to support a whole basket of renewable energy technologies, instead of focusing on just one or two technologies which are currently the most cost-effective. This point should be repeated for emphasis: one of the key ways FITs lower costs later is by producing a diversified set of technologies now. In essence, a FIT is a tool for technology development *and* cost reduction. It is one of the major advantages of FIT schemes that the technology-specific approach allows for the development of a wide range of technologies at relatively low costs. If you are planning to have a large share of renewable energies in the future electricity mix, you will need a variety of different technologies. By supporting both fluctuating technologies, e.g. wind energy and solar, and technologies that are more firm, e.g. biomass, solar thermal, geothermal, and hydroelectric, you can lay the foundation for a 100 per cent renewables-based electricity system at an early stage (see Sections 3.4 and 7.4).

Nonetheless, some regions or countries opt for supporting only one technology with a FIT. This is usually the case if additional support mechanisms are available for other technologies. A FIT for only one technology such as photovoltaics (PV), however, includes certain risks which are most of all related to public acceptance. As the electricity costs for PV are significantly higher than that of conventional energy sources and other renewable energy technologies, and the amount of electricity produced is comparatively small, the additional costs as distributed by financing mechanisms (see Section 2.7) might seem rather high to the final consumer. In contrast, if a large portfolio of technologies is eligible under the FIT legislation, the average cost for one unit of renewable electricity is rather low. To a certain extent, more mature technologies such as wind power will help less mature technologies such as PV to be developed. In this way, public acceptance can be strengthened.

When defining the technologies eligible under the FIT legislation, it is important to include precise definitions. This is especially true for biomass/waste and PV installations. The term biomass incorporates a large variety of resources, such as forestry products, animal waste, energy crops, and sometimes municipal waste. Policy makers have to decide upon the eligibility of impure biomass and waste material. Generally, the non-biodegradable fraction of waste is not eligible for tariff payment. Accordingly, the EU Directive 2001/77/EC has defined biomass as 'the biodegradable fraction of products, waste and residues from agriculture (including vegetal and animal substances), forestry and related industries, as well as the biodegradable fraction of industrial and municipal waste' (EU, 2001). In the case of PV, advanced FIT schemes differentiate between certain categories, i.e. free-standing PV installation and building-integrated PV installations (BIPV) (see Section 3.13).

For instance, the Austrian Green Electricity Act offers a typical example of a broad definition of renewable energy. All major technologies are included. In addition, impure biomass is excluded if the biodegradable fraction is below a certain percentage. The Green Electricity Act 2002, as amended in 2006, Article 5(1) 11 provides that:

> *For the purposes of this Act the term ... 'renewable energy sources' shall mean renewable non-fossil energy sources (wind, solar, geo-thermal, wave, tidal, hydropower, biomass, waste containing a high percentage of biogenous materials, landfill gas, sewage treatment plant gas and biogases).*

2.2 ELIGIBLE PLANTS

Besides eligible technologies, those designing FITs will have to determine which plants are covered under the FIT scheme. Usually, tariff payment only applies to generation plants in the given region or country. In the case of offshore wind turbines the national territory can either be limited by the UN definition of Territorial Waters, i.e. 12 nautical miles offshore, or the Exclusive Economic

Zone, i.e. 200 nautical miles offshore. However, there are some efforts to 'open up' national FIT schemes so that they can incorporate renewable electricity produced in other countries. For instance, the German parliamentarian Hans-Josef Fell suggested granting tariff payment under the German support mechanism also for electricity that could be produced in Northern Africa and then imported via high-voltage direct current transmission lines to Europe and Germany (Fell, 2009a).

Moreover, the policy maker usually limits tariff payment to the size, i.e. the installed capacity of renewable energy plants. Especially in the case of hydropower, tariff payment can be granted only to plants up to a certain maximum capacity, e.g. 20 or 100MW. The reason for this is that large scale hydropower is already slightly more competitive with conventional energy sources without any financial support, in areas with large resources. One unit of hydropower-based electricity can often be produced at costs as low as €0.02 or 0.03/kWh, whereas onshore wind and landfill gas electricity (the next cheapest sources) cost about €0.04–0.05/kWh. Besides, large-scale hydropower projects are more capital-intensive and have more significant environmental impacts than other renewables, meaning policy makers may want to consider excluding them from FIT schemes.

Some FIT schemes also apply other limitations. The Spanish FIT scheme, for instance, only grants tariff payment for installations with a maximum capacity of 50MW. These limitations are often historically grounded. In the past, it was believed that renewable energy could only cover a small share of the electricity mix and that, by definition, renewable energy power plants had to be small-scale and decentralized installations. The recent experience in many countries, however, contradicts these assumptions. Even though the decentralized application is still one of the major advantages of renewable energies, the development in wind energy shows that wind farms with several hundred megawatts of installed capacity are feasible and economically viable. Large-scale plants are also expected for other technologies, such as PV, solar thermal, geothermal and biomass. Therefore, we suggest not including plant-related capacity limits for technologies other than large-scale hydropower. Instead, tariffs should be differentiated according to the size of each plant (see Section 2.5, size-specific tariffs). Eventually, renewable energy capacity will have to replace large-scale conventional electricity plants. This requires no limits to be placed on either plant sizes or overall installed capacity (see Section 4.8). This may, however, have to be considered carefully for developing countries (see Section 5.1).

The start of generation, i.e. the moment the installation gets connected to the grid, also determines whether a plant is going to be covered by the FIT. We recommend only including newly installed capacity as old renewable power generation plants are likely to have profited from previous support instruments. Therefore, the coming into force of the legislation usually sets the starting point for eligible plants.

Theoretically, it is also possible to exclude certain producer groups from tariff payment. In the first German FIT law of 1990, for instance, the legislator decided to exclude plants where publicly owned utilities owned a significant share. This can be an appropriate step where regulators plan to liberalize electricity markets and wish to allow new actors to become competitors to well-established national utilities. However, we recommend avoiding the exclusion of any producer group from tariff payment. The open, participatory and democratic nature of FITs is one of their most important characteristics. It also, by definition, ensures that renewable energy penetration is greater as more utilities are bound by the FIT.

2.3 TARIFF CALCULATION METHODOLOGY

One of the most urgent questions for policy makers dealing with FITs is how to get the tariff level right. A tariff that is too low will not spur any investment in the field of renewable energies while a tariff that is too high might cause unnecessary profits and higher costs for the final consumer. We recommend developing a joint analytical framework for all technologies eligible under the FIT scheme in order to guarantee transparency and comparability.

In the past, regulators (and the consultants and economists they frequently employ) have applied different methodologies for tariff calculations. However, empirical evidence shows that those countries that have based their FITs on the real generation costs plus a small premium, and thus offered sufficient returns on investment, have been most successful. This approach will hence be considered as 'best practice'. Other, less successful tariff calculation methodologies are presented in Section 4.7, namely the setting of tariffs based on either the electricity price or 'avoided costs'.

Different names have been used to describe this tariff calculation approach based on actual costs and profitability for producers. The German FIT scheme is based on the notion of 'cost-covering remuneration', the Spanish support mechanism speaks of a 'reasonable rate of return,' and the French 'profitability index method' guarantees 'fair and sufficient' profitability. Despite the variety in names and notions, in all cases the legislator sets the tariff level in order to allow for a certain internal rate of return, usually between a 5 and 10 per cent return on investment per year. In some cases the rate will have to be higher as the profitability of renewable energy projects should be comparable with the expected profit from conventional electricity generation. Only if the profitability of renewable energy generation is similar to or higher than that of nuclear or fossil plants will there be an economic incentive to invest in cleaner forms of energy.

A legislator has several options to determine the tariff. As a first step, an analysis of FIT countries with similar resource conditions might be useful. Therefore, we have included many tables with data relating to real tariff levels in Chapter 6. If, for instance, the neighbouring country has a well-functioning FIT scheme, the tariffs applied in this country might serve as a point of reference. Be forewarned, though, that the mere comparison of tariff levels will not be sufficient. Many other design options which will have an impact on the profitability of a project have to be taken into account, including the duration of tariff payment, tariff degression, grid connection costs, administrative procedures and so on, each discussed below in this chapter.

After a good frame of reference is established for tariffs, cost factors related to renewable electricity generation have to be evaluated. We recommend basing the calculations on the following criteria:

- Investment costs for each plant (including material and capital costs);
- Grid-related and administrative costs (including grid connection cost, costs for the licensing procedure, etc);
- Operation and maintenance costs;
- Fuel costs (in the case of biomass and biogas); and
- Decommissioning costs (where applicable).

Based on these data, the legislator can calculate the nominal electricity production costs for each technology. Three examples are given below. Knowing the average operating hours of a standard plant and the duration of tariff payment, the legislator can fix the nominal remuneration level. For the estimate of the average generation costs, regulators can use standard investment calculation methods (such as the annuity method). The Spanish legislator even obliges renewable electricity producers to disclose all costs related to electricity generation in order to have optimal information when setting the tariff.

In the following three examples, we are going to present the German and the French approaches for tariff calculation for industrialized nations. In order to give an example of an emerging economy we are also going to elaborate on the South African approach. Moreover, we will present an online tariff calculation tool from the European PV Policy Research Platform. By definition, tariff calculation methodologies are rather technical but certainly interesting for all committed policy makers.

2.3.1 Germany: cost-covering remuneration

Under the German FIT scheme a transparent tariff calculation methodology was developed based on the electricity generation costs. This methodology is used by the German Federal Environmental Ministry (BMU) for the initial tariff proposal in the framework of the so-called Progress Report. This report is issued every four years and serves as the base for the periodical revision of the FIT scheme (see Section 2.13). It has to be noted, however, that the initial tariff proposal of the BMU is sometimes changed during the consecutive political decision-making process. In contrast to many other countries, the German FIT scheme has the legal rank of a law. Therefore, the initial proposal of the Ministry has to pass through the government and parliament and might therefore be subject to modifications.

Table 2.1 *Summary of basic data and parameters used for profitability calculation*

	Hydro-power	Biomass	Landfill, sewage and mine gas	Geo-thermal	Wind	Photo-voltaics
Imputed period under review (years)	30 /15	20	Basic case: 20 (variant for landfill gas: 6)	20 a	Basic case: 20 (variant: 16)	20
Nominal composite interest rate (%/year)	Small plants 7 Large-scale plants 8		8	8	8	Variation within sector 5–8
Inflation rate	2%/year					
Remuneration for heat (for CHP, ex-plant)	Basic case: € 25/MWh) (Variation within sector €10–40/MWh)					
Specialist personnel costs	€50,000 per person-year					
Equivalent operating hours at full capacity of electricity-led plants (hours/year)	Dependent on degree of utilization	7700	Landfill gas 7000, sewage/mine gas 7700	7700	Dependent on conditions at location	
Equivalent operating hours at full capacity of heat-led plants	–	Dependent on model case	–	–	–	–

Source: BMU, 2008

For the setting of the tariff, both the Ministry for the Economy and the Ministry for the Environment commission studies that are conducted by various independent research institutes. In addition, wide-ranging surveys on costs are conducted among producers of renewable electricity. The results are cross-checked with published cost data and empirical values from project partners of the ministries. In this way, the BMU evaluates the average generation cost of plants. To finally determine the tariff level, several basic data and parameters are compiled. Generally, tariff payment is guaranteed for 20 years. For the tariff calculation, the interest rate for capital is set at a nominal basic value of 8 per cent. This value varies slightly in the case of some technologies. The expected annual inflation is 2 per cent. The costs for

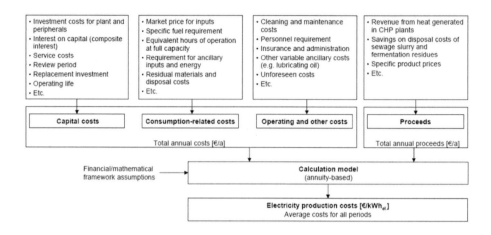

Figure 2.1 *German methodology and input variables for calculating electricity production costs*

Source: BMU, 2008

specialized personnel and the expected annual operating hours are also taken into consideration. The detailed assumptions and data are summarized in Table 2.1.

The German Ministry of the Environment applies this 'annuity method' to calculate the electricity generation costs for all renewable energy technologies except wind energy. This method of dynamic investment calculation allows for translating one-off payments and periodic payments of varying amounts into constant, annual payments (BMU, 2008). For wind power, the net present value method is applied in order to take the large variation in payment over the 20-year period into account. This variation is mostly due to a higher tariff payment in the first years of operation. All costs for renewable electricity generation are calculated on a real basis, adjusting them to inflation based on a specific reference year. Even though the German FIT is not explicitly inflation indexed (see Section 3.12), the effects are counterbalanced by the calculation method.

For the specific cost calculation, a large number of input variables have to be taken into account. This includes output data of average plants which are currently in operation, the purchasing costs for fuel in the case of biomass and biogas, investment cost (machinery, construction, grid-connection, etc), and operation costs (See Figure 2.1). Because of the unique regulatory environment in Germany, special investment costs subsidies from financial institutions are not included in the calculation, and the German FITs are based on pre-tax calculations.

2.3.2 France: profitability index method

The French government institution Agence de l'Environnement et de la Maîtrise de l'Energie (ADEME) applies what they call a 'profitability index method' for tariff calculation. It has to be noted, however, that the Ministry for the Economy, which has the final word in setting the tariff, does not apply this methodology. However, the tariff proposals of the ADEME are often translated into legislation and offer a good example of a calculation method different from Germany's. The French methodology was originally developed for wind power tariff calculation and is based on a set of transparent parameters and formulae as described below.

The profitability index (PI) is the ratio between the net present value (NPV) and the initial investment (I) of a renewable energy project. The discount rate applied in France is the average weighted cost of capital (AWCC) and not the targeted internal rate of return (IRR). For the calculation of the first French FITs in 2001, a reference value of $t = 6.5$ per cent in real terms was applied. According to the FIT payment duration, the depreciation period n was set at 15 years.

For readers that are gifted in maths and economics, the capital recovery factor (Kd) is defined by the following formula:

$$Kd = Kd(t, n) = [t(1+t)^n] / [(1+t)^n - 1] \tag{1}$$

The investment cost ratios:

> $Iu = I / P$, or $Ius = I / S$ (P and S being the rated power and the swept area) and the residual value $Valres$ of the project after n years of operation, expressed as a fraction of the initial investment.

The average, annual constant operation and maintenance (O&M) cost is expressed by the following ratio:

> $Kom = Dom / I$, where Dom stands for annual O&M costs, including repairs.

The average, annual energy yield ratio is expressed as:

> $Nh = Ey / P$ (hours per year at rated power) or as $Eys = Ey / S$ (in kWh per year and per m²) where Ey is the mean annual amount of energy sold to the grid.

For a targeted profitability index value PI the required constant tariff Teq from year 1 to year n is:

$$Teq = \{\{(1 + PI) \, Kd \, [1 - (Valres / (1+t)(n+1))] + Kom\} / Nh\} \, Iu \tag{2}$$

or the same equation using *Ius* and *Eas* in place of *Iu* and *Nh*, and for *PI* = 0, the tariff *Teq* is equal to the Overall Discounted Cost (*ODC*) of each kWh.

There is a direct link between the profitability index (*PI*) and the internal rate of return (*IRR*):

$$Kd = (IRR, n) = (1+PI) \, Kd \, (t, n)$$

For instance, if the number of years of tariff payment (*n*) is 15 and the discount rate t is 6.5 per cent, the profitability index is 0.3 and the internal rate of return is 10 per cent.

The profitability index method offers the advantage of differentiating the cost per kWh and the tariff payment. The difference between both units defines the margin for producer and therefore the profitability of a project (Chabot et al, 2002).

2.3.3 South Africa – reasonable rate of return

The South African legislator calculates their tariff based on full cost recovery and reasonable returns on investment. The National Energy Regulator of South Africa (NERSA) refers to this as the 'levelized cost of electricity' approach. In contrast to many industrialized countries, the inflation rate in developing countries or emerging economies is often higher. As we discuss in Section 3.12, this means policy makers in developing countries may want to consider inflation-indexed tariff payments. NERSA expects an annual inflation rate of 8 per cent when calculating tariffs. For financing renewable energy projects, a debt–equity ratio of 70:30 was estimated. Accordingly, the nominal cost of debt was considered to be 14.9 per cent, and the real cost of debt after tax 6.39 per cent. In contrast to the German methodology, the South African legislator also included an average tax rate of 29 per cent in the calculation. The real return on equity after tax was fixed at 17 per cent, and the weighted average cost of capital at 12 per cent.

For each technology, the following assumptions were made when calculating the tariffs. In the case of wind energy the policy makers used an average wind speed of 7 metres per second at 60 metre height to calculate the load factor. Landfill gas methane – a 'by-product' of the coal industry – is assumed to drive a gas turbine reciprocating engine. The tariff for concentrated solar power is based on the assumption of using parabolic trough plants with molten-salt storage for six hours a day. In all details, the tariff calculation for the year 2009 was based on the parameters in Table 2.2.

NERSA adjusted their FIT payment to the latest publicly available international cost and performance data of renewable energy technologies. Due to the increase in capital cost on the international financial markets, NERSA significantly increased tariff payments for almost all technologies in comparison to the first FIT proposal (see Section 6.4.2).

Table 2.2 South African methodology for calculating levelized costs

Parameter	Units	Wind	Small Hydro	Landfill gas methane	CSP
Capital cost: engineering procurement & construction (EPC)	$/kW	2000	2600	2400	4700
Land cost		5%	2%	2%	2%
Allowance for funds under construction (AFUC)		4.4%	10.6%	4.4%	4.4%
Tx/Dx Integration cost		3%	3%	3%	3%
Storage (CSP)		–	–	–	8%
Total investment cost	$/kW	2255	3020	2631	5545
Fixed O&M	2009$kW/yr	24	39	116	66
Variable O&M	2009$/kWh	0	0	0	0
Economic life	years	20	20	20	20
Weighted average cost of capital		12%	12%	12%	12%
Plant lead time	years	2	3	2	2
Fuel type		renewable	renewable	renewable	renewable
Fuel cost	$/10^6btu	0	0	1.5	0
Fuel cost	$/kWh	–	0.00106	–	–
Heat rate	Btu/kWh	–	–	13500	–
Assumed load factor		27%	50%	80%	40%
Levelized cost of electricity production	$/kWh	0.1247	0.0940	0.0896	0.2092
Exchange Rate R/S	ZAR/$	10	10	10	10
Levelized cost of electricity production	R/kWh	1.247	0.940	0.896	2.092

Note: CSP = Concentrated solar plant, parabolic trough with storage (6 hours per day). Landfill gas = methane.
Source: NERSA, 2009

2.3.4 European Photovoltaic Policy Platform online tool for tariff calculation

In view of the fact that getting the tariff right is one of the most difficult issues when drafting a FIT, the European Photovoltaic Technology Platform has developed a tool to calculate FITs for PV. It is based on several economic indicators, including system costs, systems size, annual efficiency decrease of solar PV plants, annual cost of insurance, average energy yield in a given region, and a parameter related to capital cost. The excel spreadsheet is downloadable from: www.eupvplatform.org/index.php?id=37 (scroll to the bottom of the page) and discloses all the key assumptions behind the tariff calculation model.

2.4 TECHNOLOGY-SPECIFIC TARIFFS

If the policy maker calculates the tariffs based on the generation costs of renewable electricity, technology-specific tariffs are the natural result. Technology-specific support is one of the main features of many FITs. In contrast to other quantity-based support schemes, such as tradable certificate schemes or renewable portfolio standards, FITs try to take the technology-specific generation costs into account in order to promote a broad base of different technologies (see Section 2.3, Tariff Calculation Methodology). Technology-specific support is necessary because of the large differences in generation costs among renewable energy technologies. While certain types of biomass or biogas can already be produced for less than €0.03/kWh, less mature technologies such as photovoltaics are produced at a cost of more than €0.43/kWh (Ragwitz et al, 2007).

Further differentiation might be necessary within the generic group of biomass products. As mentioned above, biomass fuel types include forestry products, animal waste, energy crops, and sometimes waste or the biodegradable fraction of waste. Generation costs vary widely as, for instance, energy crops are generally more expensive than residues from forestry, and producing biogas from animal residues is more expensive than the generation of landfill or sewage gas. Therefore, some FIT schemes take different fuel types for biomass plants into account (see the tables for remuneration in Spain and Germany, Sections 6.1.1 and 6.1.2). In addition, the cost for different transformation processes of biomass to electricity, e.g. co-combustion and gasification, might have to be reflected in the tariff design.

2.5 SIZE-SPECIFIC TARIFFS

Besides technology-specific tariffs, many FIT schemes include different remuneration levels for different sizes of a given technology. The underlying idea is that larger plants are generally less expensive. Therefore, most FIT schemes set specific tariffs for a particular technology in relation to plant size. The easiest way is to establish different groups according to the installed capacity, e.g.:

- 0kW < Tariff/Price ≤ 30kW;
- 30kW < Tariff/Price ≤ 100kW;
- 100kW < Tariff/Price < 2MW;
- 2MW and above.

The choice for the range of each group does not necessarily have to be random. Many technologies offer standard products of a certain size range. In the case of PV, for instance, a typical rooftop installation for private households has a capacity of 3–30kW. Larger-scale rooftop installations for industrial buildings or farms usually have an installed capacity of up to 100kW. Therefore, an analysis of standard products of a certain technology in a given region or country will help to set plant-size-specific tariffs. In order to avoid potential disruptive effects through size categories, the legislator also has the option to develop a formula which relates the plant size to the tariff payment.

2.6 DURATION OF TARIFF PAYMENT

The duration of the tariff payment is closely related to the level of tariff payment. If a legislator desires a rather short period of guaranteed tariff payment, the tariff level has to be higher in order to assure the amortization of costs. If tariff payment is granted for a longer period, the level of remuneration can be reduced. However, in the case of longer payments inflation will be greater and must be factored in (see Section 3.12). FITs around the world usually guarantee tariff payment for a period of 10–20 years, while a period of 15–20 years is the most common and successful approach. A payment of 20 years equals the average lifetime of many renewable energy plants. Longer remuneration periods are normally avoided because otherwise technological innovation might be hampered. Once tariff payment ends, the producer will have a stronger incentive to reinvest in new and more efficient technologies instead of running the old plant in order to receive tariff payment. However, producers normally have the right to continue selling electricity under standard market conditions.

Accordingly, the guidelines for the South African FIT state:

> 7.6. *Following the completion and end of the duration of the contracted REFIT tariff, the Generator shall be required to negotiate tariffs under market conditions applying at the time.*

When fixing the duration of tariff payment, policy makers should clearly state whether producers have the right to leave the FIT scheme during the guaranteed payment period. This might be of interest for renewable electricity producers if the spot market power price for 'grey' electricity, i.e. fossil-based or nuclear power, rises above the guaranteed FIT. In countries which have started to incorporate the negative external costs of fossil fuels and remove subsidies for conventionally produced electricity, this will probably start to occur more in coming years, especially for the most cost-effective renewable energy technologies such as wind energy and landfill gas capture.

In this case, the legislators basically have three options:

1 They can mandate that the FIT duration period has to be 'fulfilled' and the renewable electricity producer does not have the right to enter the 'grey' power market. The positive effects of this approach are the lower electricity costs for final consumers, once the power price for conventional power exceeds the guaranteed tariff level. In this case, the FIT will stabilize and lower the average electricity price. However, such a policy could delay the integration of green electricity into the grey power market as developers will be getting less for their renewable electricity.
2 Regulators can state that the renewable electricity producer has the right to leave the FIT but no right to re-enter the FIT scheme. This would in essence complicate the participation of renewable electricity producers in the conventional grey market as future prices might be difficult to anticipate.
3 The legislation can give the producer the opportunity to switch between the guaranteed remuneration under the FIT and the participation within the spot market for electricity. By those means, the producer can gather first-hand experience in the power market without being exposed to all risks related to volatile market prices. In this case, regulators would determine a time period in which the producer is allowed to change between both systems, e.g. once every month or once every year. This approach is presented in more detail in Section 3.1.

2.7 Financing Mechanisms

Another basic feature of FITs is that additional costs are distributed equally among all electricity consumers. This financial burden-sharing mechanism permits the support of large shares of renewable electricity with only a marginal increase of the final consumer's electricity bill (and, if done properly, FITs end up saving consumers money as they displace fossil fuels and reduce greenhouse gas emissions). Moreover, by determining tariff payment and establishing the purchase obligation, the national government only acts as a regulator of private actors in the electricity market. No government financing is included under these conditions. Alternative financing mechanisms have proven to be too sensitive towards external effects, such as changes in government or general economic downturns (see Section 4.6).

In order to pass the price from the producer of renewable electricity to the final consumer, the costs, i.e. the aggregated tariff payments, must be passed along the electricity supply chain. First, the producer of renewable electricity receives the tariff payment from his or her local grid operator. By legal obligation through the FIT scheme, this grid operator is obliged to pay, connect and transmit the produced electricity (see Section 2.8). Normally, renewable electricity producers get connected to the next distribution system operator (DSO). In some cases, however, a producer of a large plant might also decide to connect directly to higher voltage lines, through the transmission system operator (TSO). Afterwards, the costs and the accounting data are passed to the next highest level in the electricity system until the national TSO aggregates all costs and divides it by the total amount of renewable electricity produced.

From this stage, regulators can choose two alternative approaches. The costs can be equally distributed among all national supply companies in relation to the total amount of electricity provided to the final consumer. This way, all final consumers pay the same for the total amount of renewable electricity produced in a given territory. Alternatively, the DSO or TSO can also distribute all costs by a small increase in the pass-through cost of the electricity grid (see Figure 2.2).

In some countries, exceptions from this equal burden-sharing mechanism have been made for energy-intensive industries (see Section 3.14).

2.8 Purchase Obligations

Besides long-term tariff payments, the purchase obligation is the second most important 'ingredient' for all FIT schemes as it assures investment security. It

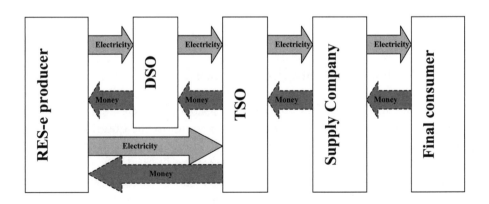

Figure 2.2 *General flow of electricity and financing under FIT schemes*

Note: RES-e = electricity from renewable energy sources.
Source: Jacobs, 2009

obliges the nearest grid operator to purchase and distribute all electricity that is produced by renewable energy sources, independent of power demand. This means, for instance, that in times of low demand, the grid operator will reduce the amount of 'grey' electricity while all 'green' electricity is incorporated into the electricity mix. The purchase obligation is especially important for more variable renewable energy technologies, such as wind and solar PV, as the producer cannot control when the electricity will be generated. In contrast, gas- and coal-fired and nuclear-based power plants can increase and reduce output, along with hydroelectric dams, biomass facilities and geothermal power stations. Therefore, advanced FIT schemes sometimes include tariff differentiation according to electricity demand (see Section 3.3, Demand-Oriented Tariff Differentiation).

The purchase obligation protects renewable electricity producers in monopolistic or oligopolistic markets where the grid operator might also dispatch power generation capacity. When decisions are made about which power generation sources to use to meet electricity demand, such grid operators might be biased and dispatch power from power plants such as their own first. (It should be noted that sometimes the purchase obligation does not apply under premium FITs; see Section 3.1.)

As an example of a well-designed purchase obligation, the German FIT establishes an obligation to purchase, transmit and distribute all electricity produced under the FIT scheme.

Section 8, paragraph 1 of the Renewable Energy Sources Act (EEG) 2009 states:

> *grid system operators shall immediately and as a priority purchase, transmit and distribute the entire available quantity of electricity from renewable energy sources and from mine gas* (BGB, 2008)

2.9 PRIORITY GRID ACCESS

Unfair grid access rules are often a barrier in power markets where the grid operator itself is engaged in power production. This lack of 'unbundling' generation, transmission and distribution might lead to a situation where the grid operator prioritizes its own generation units when it comes to the question of which power plant will get connected to the grid. Therefore, FITs usually include provisions that eligible plants must be connected to the grid. The German FIT scheme, for instance, states that 'grid system operators shall immediately and as a priority connect plants generating electricity from renewable energy sources'. We recommend this approach as the 'immediate' connection prevents delays by the grid operator and 'priority' connection enables renewable energy plants to get connected to the grid before conventional power generation units.

Equally, the lack of transmission capacity can seriously offset the deployment of renewable energies. In Ontario, for instance, the FIT of 2006 even states that:

> *Applicants are cautioned that certain areas of the transmission grid are limited in their ability to accept incremental power. For this reason, the OPA [Ontario Power Authority] may be required to restrict or decline project applications in certain designated areas.*

Instead of taking bottlenecks in the grid for granted and refusing green electricity producers to access the grid, the legislators should follow the German model, and not the Canadian, to initiate grid reinforcement and the establishment of national grid extension plans.

2.10 Cost-Sharing Methodology for Grid Connection

Grid connection rules impact the overall profitability, and therefore success, of renewable energy support policies. Even though other support mechanisms may be well established in a given country, discriminatory practices, regulations, interconnection standards and other rules might offset or seriously disturb the deployment of renewable energy projects. This is due in particular to the high costs for grid connection in relation to the total project costs. The European study *GreenNet-Europe* has calculated that, in the case of offshore wind power plants, grid connection can account for up to 26.4 per cent of total investment costs.

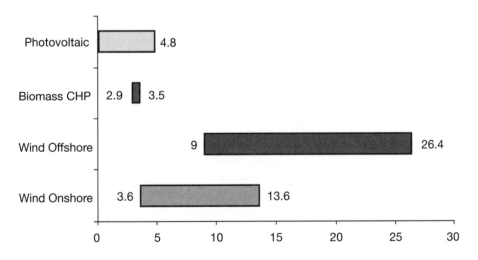

Figure 2.3 *Percentage of grid integration costs compared to total investments*

Source: GreenNet-Europe, undated

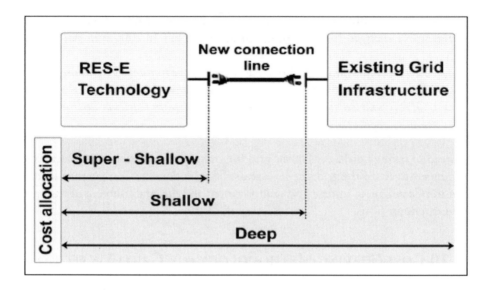

Figure 2.4 *Cost-sharing methodology for grid connection*

Source: Auer et al, 2007

Even though the share is lower for all other renewable energy technologies, the methodology for cost sharing of grid connection is often essential when it comes to the decision as to whether a project is profitable or not (See Figure 2.3).
Many FITs usually define the methodology that is used for dividing the costs for grid connection between the renewable electricity producer and the grid operator. Some legislators prefer to establish these rules in legislation for grid regulation. Essentially, three different methodologies can be applied to connection charging: the 'deep', the 'shallow' and the 'super-shallow' (Knight et al, 2005) (See Figure 2.4).

The so-called deep connection charging approach leaves the producer of renewable electricity with all costs, both for grid connection and for grid reinforcement. This includes the costs for the connection line to the next connection point as well as the costs for reinforcing the already established grid infrastructure. In the case of a lack of transmission capacity, the producer is obliged to pay for the necessary upgrading. We do not recommend this approach. Historically, it was employed for large-scale conventional power plants. In the light of the high investments costs for these power plants, the additional expenditures for grid connection under the deep approach were negligible. This is different for renewable energy projects, which tend to have much lower overall costs per project than mammoth nuclear and coal-fired units. Furthermore, the deep approach provides an incentive to produce electricity in areas with a well-developed electricity grid.

This makes sense in the case of coal- or gas-fired power plants but not in the case of renewable energy projects. Wind power plants, for instance, should be built in the windiest locations and not just in regions with available grid capacity.

As an alternative, the shallow connection charging approach was developed. It states that the renewable energy producer only has to pay for the new electricity line to the next grid connection point, while the grid operator has to cover all costs for potential reinforcement of existing grid infrastructure. The costs covered by the grid operator will be passed on to the final consumer in terms of system charges. Under this approach, the renewable electricity producer will choose the location for the power plant depending on the resource availability (e.g. wind speed) and not infrastructure availability.

Accordingly, the German EEG of 2009 (Chapter 3, Sections 13 and 14) states that:

> *The costs associated with connecting installations generating electricity from renewable energy sources or from mine gas to the grid connection point ... shall be borne by the installation operator. The grid system operator shall bear the costs of optimizing, boosting and expanding the grid system.*

It is also possible to mix both approaches. In this case, the power producer will pay for the electricity line to the next connection point. The costs for grid reinforcement will be shared between the grid operator and the electricity producer. Normally, the share covered by the producer depends upon the assessment of their proportional use of new infrastructure. This combination can be seen as a compromise between an incentive for using available grid infrastructure and choosing the resource optimal locations.

Recently, a super-shallow connection charging approach was implemented in some European countries to promote the deployment of offshore wind power plants, particularly in Denmark and Germany. Connection lines from offshore wind fields to the nearest onshore connection point are rather expensive because of the long distances involved. To free the offshore wind power developers from this financial burden, legislators decided that even the costs for the new connection line from the offshore wind park to the next onshore connection point have to be paid by the grid operator.

We recommend using the shallow grid connection approach or even the super-shallow grid connection approach. This allows for a strict separation of infrastructure investment and investment into new generation capacity. There is clearly a tendency for countries wanting to promote renewable energies to move away from the deep to the shallow connection charging approach. Whatever cost-sharing methodology regulators wish to apply, they must take the financial advantages (super-shallow approach) or disadvantages (deep connection charging approach) of green electricity producers into account when calculating the tariffs.

The estimated costs for grid connection and reinforcement must be part of the tariff calculation methodology (see Section 2.3).

2.11 Effective Administrative Procedures

The experience of some FIT countries shows that, despite good economic and grid access conditions, generation capacity for renewable electricity does not increase significantly. The reasons for mediocre performance despite having the best designed FIT can include administrative barriers such as long lead times for project approval, a high number of involved authorities and the lack of inclusion into spatial planning (Ragwitz et al, 2007; Roderick et al, 2007; Coenraads et al, 2008). The European Commission, for example, recommends implementing quicker permitting procedures for small-scale projects because they differ fundamentally from large-scale coal-fired power plants (EU Commission, 2005). It makes little sense to force both types of projects to go through the same permitting process. The most nefarious administrative barriers appear related to lead times, coordination and spatial planning.

2.11.1 Minimizing lead times

One major administrative barrier for renewable energy projects is long lead times. In the EU, lead-times for small-scale hydropower development vary from 12 months (Austria) up to 12 years (Portugal and Spain). In France, wind power developers sometimes have to wait for four to five years to move from a project outline to electricity production. The same is the case for offshore wind power projects in Ireland.

Policy makers can reduce this barrier by establishing a time limit on the entire permitting process of all necessary documents from all organizations that are involved. National and local entities will be forced to deal with project permissions in time and organizations opposed to renewable energies will have less influence when it comes to non-economic barriers. Setting deadlines for the decisions of each authority will help, as long as authorities can keep to them. Especially on a local level, administrative bodies often lack experience in dealing with industrial size projects.

2.11.2 Minimizing and coordinating the authorities involved

Another important constraint for the development of renewable energies is the large number of authorities involved in the licensing process. In France, for instance, wind power producers will have to get in contact with 27 different authorities at different political levels. In some Italian regions, up to 58 permits from different authorities are needed for small-scale hydropower plants.

Complexity can be reduced by clarifying the responsibilities of each authority, and establishing a new organization dedicated to rapid renewable energy deployment, sometimes called a 'bottleneck organization' or 'one-stop shop', to coordinate and simplify the planning process. Most successful are those countries that authorize one single administrative body to deal with all subordinated authorities at different political levels. This approach was proposed by the European Commission in the framework of the new Directive for Renewable Energies.

2.11.3 Inclusion in spatial planning

Spatial planning provisions help to organize the use of physical space in a given region or country, such as stipulating where roads, industrial areas, power plants and sewer systems should be located. Spatial planning at local level must anticipate future renewable energy projects by including them when drafting or revising regulations and standards. In this process the available resources, such as wind speed and solar radiation, should be identified. The German building code of 1996, for instance, obliged each community to designate specific areas for the development of wind power projects. By those means, the legislator managed to shorten the administrative process considerably.

2.12 Setting Targets

Sometimes FIT legislation is combined with ambitious political targets for renewable energies. This has merit, as targets are important in signalling long-term political commitment to investors. They indicate that support mechanisms will remain in place for a certain period of time and they increase the likelihood of tariffs being sufficiently high. Targets should always be formulated as minimums by including the words 'at least' (e.g. 'at least 20 per cent by 2020'). This way, targets do not have the negative effects of acting as capacity caps (see Section 4.8), where the deployment of new installation slows or comes to a halt once the target has been reached.

Targets can be formulated as a certain share of renewable energies in the overall energy or electricity mix. This has been done by the German legislator who determined that the German FIT scheme 'aims to increase the share of renewable energy sources in electricity supply to at least 30 per cent by the year 2020 and to continuously increase that share thereafter'. Alternatively, targets can also be established for the installed capacity. The policy makers in Ontario combined both issues. In 2004, the government decided to set a target of '5 per cent (1350MW) of its electricity from renewable sources by 2007 and 10 per cent (2700MW) by 2010'. We recommend establishing targets for the short, mid and long term, thus establishing a pathway of how renewable energies can increasingly substitute fossil and nuclear power generation sources. In order to determine to what extent those targets have been achieved, we also recommend including a progress report.

2.13 Progress Reports

Last, but not least, evaluating and periodically reporting on the state and progress of FIT programmes is crucial for long-term success. Reporting and evaluation is usually the task of the responsible ministry. They will ensure that the law is functioning well, and if necessary, recommend how it could be improved or amended. In some countries, progress reports provide the scientific grounds for periodic amendments of FIT schemes. The Spanish and German FIT schemes, for instance, are modified every four years. This periodic revision guarantees stability for the producers, who know that the legislation will not be changed in the meantime, but it also gives politicians room for modifications. When regulators implement a FIT scheme for the first time, frequent adjustments might be necessary in the first couple of years. Therefore, the South African legislator has chosen to review the FIT scheme every year for the first five years and afterwards only every three years.

Progress reports typically include an analysis of the growth rates and the average generation costs of all eligible technologies. They identify the economic, social and environmental costs and benefits of renewable energy support (especially an estimate of greenhouse gas reductions). They review the additional costs for the final consumer. And they calculate the ecological effects of renewable energy plants, positive and negative, on nature and landscape.

2.14 Checklist for a Basic FIT Scheme

To summarize this chapter, we have developed a checklist that regulators (and anyone with an interest) can refer to when drafting a basic FIT scheme:

- Choose the eligible technologies based on the resource availability in your country.
- Determine which kind of power production plants shall be eligible.
- Establish a transparent tariff calculation methodology based on the generation costs of each technology.
- Set technology- and size-specific FITs.
- Fix the duration of tariff payment (usually 20 years).
- Create a robust financing mechanism, sharing the additional costs among all electricity consumers.
- Oblige the grid operator to purchase all renewable electricity.
- Grant priority grid access.

- Regulate the cost sharing for grid connection and reinforcement based on the 'shallow' or 'super-shallow' connection charging approach.
- Create effective administrative procedures.
- Set renewable energy targets and mention them explicitly in the FIT legislation.
- Establish a progress report as the scientific basis for future adjustments.

REFERENCES

Auer, H., Obersteiner, C., Prüggler, C., Weissensteiner, L., Faber, T. and Resch, G. (2007) *Action Plan: Guiding a Least Cost Grid Integration of RES-Electricity in an Extended Europe*, GreenNet-Europe, Vienna, May

BGB (2008) 'Gesetz zur Neuregelung des Rechts der erneuerbaren Energien im Strombereich und zur Änderung damit zusammenhängender Vorschriften', *Bundesgesetzblatt*, vol 1, no 49, p2074

BMU (2008) *Depiction of the Methodological Approaches to Calculate the Costs of Electricity Generation Used in the Scientific Background Reports Serving as the Basis for the Renewable Energy Sources Act (EEG) Progress Report 2007*, Extract from Renewable Energy Sources Act (EEG), Progress Report 2007, Chapter 15.1

Chabot, B., Kellet, P. and Saulnier, B. (2002) *Defining Advanced Wind Energy Tariffs Systems to Specific Locations and Applications: Lessons from the French Tariff System and Examples*, paper for the Global Wind Power Conference, April 2002, Paris

Coenraads, R., Reece, G., Voogt, M., Ragwitz, M., Resch, G., Faber, T., Haas, R., Konstantinaviciute, I., Krivosik, J. and Chadim, T. (2008) *Progress: Promotion and Growth of Renewable Energy Sources and Systems, Final Report*, Contract no. TREN/D1/42-2005/S07.56988, Utrecht, Netherlands

EU (2001) 'Directive 2001/77/EC of the European Parliament and of the Council of 27 September 2001 on the promotion of electricity produced from renewable energy sources in the internal electricity market', *Official Journal of the European Communities*, L 283/33

EU Commission (2005) *The Support of Electricity from Renewable Energy Sources*, Communication from the Commission, COM(2005) 627 final, Brussels

Fell, H.-J. (2009a) *Mögliche Gesetzentwicklung zur Realisierung erster Leitungen und Kraftwerke in Nordafrika*, Presentation at the Official Presentation of the Desertec Foundation, 17 March, Berlin

Fell, H.-J. (2009b) *FIT for Renewable Energies: An Effective Stimulus Package without New Public Borrowing*, www.boell.org/docs/EEG%20Papier%20engl_fin_m%C3%A4rz09.pdf

Grace, R., Rickerson, W. and Corfee, K. (2008) *California FIT Design and Policy Options*, CEC-300-2008-009D, California Energy Commission, Oakland, CA

GreenNet-Europe (undated) http://greennet.i-generation.at/

Jacobs, D. (2009) *Renewable Energy Toolkit: Promotion Strategies in Africa*, World Future Council, May

Klein, A., Held, A., Ragwitz, M., Resch, G. and Faber, T. (2008) *Evaluation of Different FIT Design Options*, best practice paper for the international Feed-in Cooperation, www.feed-in-cooperation.org/images/files/best_practice_paper_2nd_edition_final. pdf

Knight, R. C., Montez, J. P., Knecht, F. and Bouquet, T. (2005) *Distributed Generation Connection Charging within the European Union – Review of Current Practices, Future Options and European Policy Recommendations*, www.iee-library.eu/images/all_ ieelibrary_docs/elep_dg_connection_charging.pdf

Mendonça, M. (2007) *Feed-in Tariffs: Accelerating the Deployment of Renewable Energy*, Earthscan, London

NERSA (2009) *South African Renewable Energy FIT (REFIT), Regulatory Guidelines*, National Energy Regulator of South Africa, www.nersa.org.za/UploadedFiles/ ElectricityDocuments/REFIT%20Guidelines.pdf, 26 March 2009

Ragwitz, M., Held, A., Resch, G., Faber T., Haas, R., Huber, C., Coenraads R., Voogt, M., Reece, G., Morthorst, P. E., Jensen, S. G., Konstantinaviciute, I. and Heyder B. (2007) *Assessment and Optimisation of Renewable Energy Support Schemes in the European Electricity Market*, final report, OPTRES, Karlsruhe

Roderick, P., Jacobs, D., Gleason, J. and Bristow, D. (2007) *Policy Action on Climate Toolkit*, www.onlinepact.org/, accessed 23 May 2009

Sösemann, F. (2007) *EEG – The Renewable Energy Sources Act: The Success Story of Sustainable Policies for Germany*, Federal Ministry for the Environment, Nature Conservation and Nuclear Safety, Berlin

3
Advanced FIT Design Options

This chapter explores advanced FIT designs for those countries where FITs have been operating for some length of time and/or a sufficiently robust and sizeable renewable energy industry has been established. As the share of renewable electricity increases, additional measures will have to be taken to avoid excess profits and costs. At the same time, the incorporation of 'green' electricity into the traditional 'grey' power market becomes increasingly important. This process can be enhanced by advanced design options for better market integration.

More specifically, this chapter will discuss advanced FIT design options that fall into three categories. First, it will explore those aiming for better market integration, such as:

- premium FITs;
- incentives for participation in the conventional energy market;
- demand-oriented tariff differentiation;
- tariff payment for combining technologies;
- forecast obligation; and
- special tariff payment for grid services.

Second, it will investigate FITs designed to ensure economic efficiency and keep windfall profits to a minimum, such as:

- location-specific design;
- tariff payment for limited or total electricity generation;
- tariff degression; and
- flexible tariff degression.

Third, other 'novel' design options will be explored, such as:

- increasing tariffs;
- inflation-indexed tariffs;
- additional tariff payment for innovative features; and
- exclusion for energy-intensive industries.

It is important to note that the choice and implementation of these advanced options depends on particular national markets for electricity as well as other macroeconomic data. With the implementation of each advanced design option, policy makers will have to determine whether the expected benefits will outweigh the additional complexity of the FIT scheme.

3.1 Premium FITs

For better integration of renewable electricity into the conventional energy market, more and more countries have chosen to implement so-called 'premium' FITs. This can also help to win the support of, or neutralize criticism from, those ideologically opposed to FITs, as their general critique of price-based support instruments as being non-market-based will be weakened. The premium, or 'market-dependent' tariff (Couture, 2009) was first implemented by Spain in 1998, followed by Denmark, Slovenia, the Czech Republic, Italy, Estonia, Argentina and the Netherlands. Finland also plans to implement this remuneration option and Germany is still discussing it.

Under a premium FIT, the remuneration for a green electricity producer consists of two elements: the general, hourly market price of electricity on the conventional power markets, and a reduced FIT which has to be sufficiently high to allow for reasonable profitability of renewable energy projects. This combination of two remuneration components makes it harder for the legislator to fix the FIT for each technology as market prices are fluctuating. The development of the market price has to be anticipated in order to avoid windfall profits (in the case of high market prices) and guarantee sufficient returns of investment (in the case of low market prices).

The Spanish legislator found that anticipating market prices is extremely difficult. Therefore, in 2007 the Spanish premium FIT scheme was complemented by a 'cap' and a 'floor'. The cap impedes the combined remuneration from exceeding a certain limit in the case of high market prices, while the floor prevents the remuneration falling below a minimum threshold in the case of low market prices. For onshore wind power, for instance, the premium was fixed at €0.029/kWh. The combined tariff payment (market price plus premium tariff) cannot exceed €0.085/kWh and cannot fall under €0.071/kWh (See Figure 3.1). We consider the implementation of a cap and a floor to be crucial for effective and efficient support.

There are certain prerequisites which will help to guarantee a smooth implementation of premium FITs:

- Effective ownership unbundling (the strict separation of power producers and grid operators) will help to guarantee fair grid access for renewable energy producers. Under this remuneration option, green power producers can no

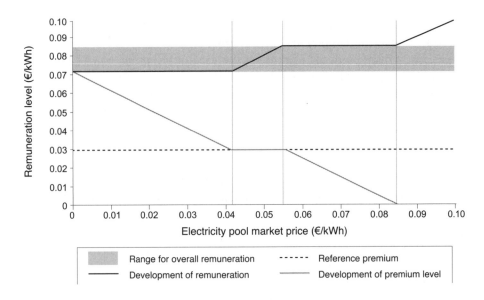

Figure 3.1 *Progression of the remuneration level under the Spanish premium FIT (wind onshore)*

Source: Klein et al, 2008

longer rely on a purchase obligation (see Section 2.8). Therefore it is important that the grid system operator is not biased when dispatching power units.
- Liquidity on the conventional power markets will lead to stable and less volatile prices. Therefore, it is important that a large share of overall electricity is traded on the spot market and not via bilateral agreements. This will make it easier for the legislator to anticipate power prices.
- The access conditions to the conventional power market will have to be adjusted to the needs of renewable energy technologies. It is especially important to establish several intra-day markets, as the power generation from certain renewable energy technologies, e.g. wind and solar power, is hard to predict a day ahead. Intra-day markets enable market participants to sell their power several times a day and not only one day ahead, which suits producers of fluctuating power better. The Spanish regulation even allows wind power producers to make hourly adjustments to the power production forecasts (see Section 3.5).

Due to the fact that one part of the remuneration – the market price – is volatile, the investment security for the renewable electricity producer is lower than in the case of the normal fixed tariff option. This risk factor makes renewable energy projects slightly more expensive, as banks normally demand higher interest rates to compensate for such insecurity. Therefore, when calculating the tariffs, the expected return on investment has to be slightly higher under the premium FIT option. To give an example, the Spanish legislator calculated the tariffs based on 7 per cent returns on investment under the fixed tariff option, and 5–9 per cent under the premium FIT option.

We strongly recommend implementing the premium FIT only as an alternative option to the fixed FIT, and not a substitute for it. Without the alternative of a fixed tariff option, you would exclude many potential renewable electricity producers from the support scheme, especially private persons and small and medium companies. As a matter of fact, a private person with a small PV system on the roof will not be willing to engage in selling electricity on the spot market with all the transaction costs that would be included. The same might be the case for community wind farms. On the contrary, large players and especially utilities are already experienced when it comes to selling electricity on the market. They will almost certainly opt for this remuneration option, especially when the expected rate of return is slightly higher.

The combination of two remuneration options forces the policy maker to decide to what extent producers will be allowed to switch between both options. In Spain, producers have the chance to switch between both remuneration options on an annual basis. In Germany, a monthly change has been envisaged by the legislator. The policy maker should be careful when deciding upon these changes, as periods that are too short might incentivize 'cherry-picking'. In times with high market prices, producers will opt for the premium FIT, while in times of low market prices everyone changes back to the fixed tariff option.

Ultimately, the policy maker will have to decide on whether the expected advantages of a premium FIT outweigh the disadvantages. The positive effects are certainly better market integration and the opportunity for green energy producers to gather experience with selling electricity on the market. Besides, producers are still able to get relatively good financing conditions for projects as the bank usually takes the fixed FIT as a basis. This, of course, is only true for countries where the producer can choose between both remuneration options. The negative drawbacks are higher costs for the final consumer due to less investment security, and higher administrative costs because the setting of the tariff, the cap and the floor are more complex.

3.2 Incentives for Participation in the Conventional Energy Market

In order to smooth the path from the fixed tariff payment to participation in the conventional energy sectors, legislators can offer an additional incentive for selling electricity on the spot market. Spain followed this approach from 2004 until 2007. At that time, the tariff payment was still expressed as a share of the electricity price. In the case of wind power, for instance, producers were offered 40 per cent of the average electricity price under the premium FIT option. In addition, producers of renewable electricity received an extra 10 per cent for market participation. According to the Spanish wind industry association, the latter amounted to an additional tariff payment of €0.007/kWh (AEE, 2008). This incentive can help to compensate for the additional transaction costs related to the activities on the market.

3.3 Demand-Oriented Tariff Differentiation

Recently, several countries have chosen to differentiate tariff payment according to electricity demand. The policy maker can decide to pay a higher tariff in times of high demand (peak) and lower tariffs in time of low demand (off-peak). This way, producers will have an incentive to align their production to general consumption patterns.

The Hungarian FIT scheme constitutes an interesting example, even though the support mechanism has often been criticized for not sufficiently differentiating technologies (in the table below you can see that wind power and solar power receive the same tariff). However, it is interesting to see that wind and solar power are exempt from demand-oriented tariff differentiation, as producers of wind power and solar PV cannot control the timing for electricity consumption. As shown in Table 3.1, all other technologies receive differentiated tariff payments during peak (morning and evening), off-peak (often midday) and deep off-peak periods (night time).

Table 3.1 *Demand-oriented tariff payment under the Hungarian FIT scheme*

Technology	Tariff level (€/kWh)		
	Peak	Off-peak	Deep off-peak
Solar PV, wind	0.105	0.105	0.105
Geothermal, biomass, biogas, small hydro (less than 5MW)	0.1173	0.105	0.0429
Hydro (more than 5MW)	0.073	0.0467	0.0467

Source: Klein et al, 2008

3.4 Tariff Payment for Combining Technologies

As the share of renewable electricity increases, it will become more and more important to improve the 'quality' of the green electricity, i.e. the ability for renewable supply to match demand. In order to guarantee steady and demand-oriented electricity supply in the future it will be necessary to combine different renewable energy technologies (see Section 7.4). FITs could offer additional financial incentives to foster this combination of technologies. Accordingly, the German Renewable Energy Federation (BEE) proposed the so-called '4000 full load hours approach' or 'integration bonus'.

On average, wind power plants in Germany operate for 2000 full load hours annually. Through the combination with other technologies, such as biomass, hydropower or other storage technologies, the total amount of full load can be doubled and, more importantly, electricity output can be planned. According to the calculations of the BEE, an additional tariff of €0.02/kWh could cover the related costs. Until very recently, the German Grand Coalition of the Conservative Party and the Social Democrats debated this approach in combination with the premium FIT option. The German legislator wanted to give producers the opportunity to choose between both options. Unfortunately, the political negotiations between the two major parties had stopped at the time of writing. Therefore, no such regulation will be implemented in Germany before the general elections in September 2009.

Nonetheless, we believe that this proposal might be interesting for policy makers all over the world as it offers an additional method of improving the market integration of renewables. In addition, it demonstrates the enormous flexibility of FITs.

3.5 Forecast Obligation

The larger the share of renewable electricity becomes, the more important it is for the system operator to know how much power will be provided at a given moment in time. All conventional power plants are obliged to forecast their production. Power can be traded based on long-term bilateral agreements, on the spot market, balancing markets, or on the forward markets, and in different time horizons (month ahead, week ahead, day ahead, intra-day, etc). In some countries, wind power production has already reached a significant share, so that power forecast is increasingly important. Power prediction has not yet become a big issue with solar PV as the share of electricity has remained relatively small.

Similar to the premium FIT option, Spain was the international trendsetter for power forecast requirements. The current FIT legislation, Royal Decree 661/2007, established forecast obligation for both the fixed tariff option and the premium option. For the latter, the general rules for all electricity producers apply. In the case of the fixed tariff option, in 2004 all wind power parks above 10MW were obliged to provide forecasts. This threshold was reduced to 5MW in 2007 and 1MW by 2009. At present, all wind power producers must forecast their electricity production. Forecasts have to be made at least 30 hours beforehand, with the possibility of adjustment one hour before. If the actual power production deviates more than 5 per cent from the hourly forecast, penalties must be paid.

Prediction errors for wind power can be reduced by using shorter prediction periods (through the establishment of intra-day markets and the possibility of hourly adjustments) and the extension of the area for which predictions will be made. Therefore, it might not be the best idea to obligate each wind power producer to make production estimates for the following day. Instead, forecast for wind power production can be made by the local grid operator at regional level, as in the case of Germany. If the legislator does choose to oblige each single producer, the forecasts in one area can be aggregated, thus lowering the potential for forecast errors.

Besides, modern wind power plants often include electronic equipment allowing them to be remote-controlled. In this case, the grid operator can decrease or increase the power output of a wind turbine from a national or regional control centre. The Spanish legislator has obliged all wind power producers with more than 10MW capacity to be retrofitted in order to profit from tariff payment. The new German FIT scheme even requires that all installations with more than 100kW capacity can reduce power output via remote control. For old installations, the deadline for this requirement is 2011. These stricter conditions are due to the fact that by 2020 Germany will produce about 30–35 per cent of its electricity from renewable energy sources. In order to ensure the stability of the grid, even small production units need to have the capacity to regulate power output.

If plants are remote-controlled, it must be ensured that the grid operator only makes use of this possibility when facing grid overload or other severe problems. In Section 11 of the EEG 2009, the German legislator clarified that:

> *grid system operators shall be entitled, by way of exception, to take technical control over installations connected to their grid system with a capacity of over 100 kilowatts for the generation of electricity from renewable energy sources, combined heat and power generation or mine gas, if:*
>
> 1 *the grid capacity in the respective grid system area would otherwise be overloaded on account of that electricity;*
> 2 *they have ensured that the largest possible quantity of electricity from renewable energy sources and from combined heat and power generation is being purchased;*
> 3 *they have called up the data on the current feed-in situation in the relevant region of the grid system.*

In Germany, renewable electricity producers whose power output has been reduced by the grid operator can demand financial compensation. Compensation is mandatory and grid operators even have to state the reason why the power output has been reduced.

3.6 Special Tariff Payment for Grid Services

Nowadays, wind power plants can provide important grid services, including the capacity to support voltage dips and the provision of reactive power for grid stability. The legislators in Spain and Germany have decided to grant a special tariff payment for these services for a number of years, as retrofitting of old wind turbines is relatively costly.

Formerly, wind power plants had to be disconnected when so-called voltage dips occurred. These are short reductions of the alternate current mains voltage, normally no longer than a few seconds. When wind power plants got disconnected it normally took some time to restart and reconnect them. Today, modern technical equipment allows wind power plants to stay connected, thus supporting the stability of the grid. In 2004, the Spanish legislator granted an extra tariff payment of 5 per cent of the average electricity price. This extra remuneration was only granted for the first four years of operation. The producers had to provide a certificate that their wind turbines were capable of supporting voltage dips. In 2007, the special tariff was fixed at €0.0038/kWh for all installations that were connected before January 2008 (Jacobs, 2008). New installations are expected to provide these services without any extra payment. In the case of reactive energy, the Spanish legislator established a 'bonus–malus' system, rewarding producers which fulfil the requirements of the grid operator and penalizing producers which do not comply.

The installed wind capacity will probably reach 36,000MW by 2020. Consequently, wind power can play an important role when it comes to auxiliary grid services. Inspired by the Spanish FIT scheme, the German legislator granted a special tariff payment of €0.005/kWh for wind energy producers as long as they comply with national system service regulation. This so-called 'system service bonus' is only paid for wind power plants that are installed before 2014 and is limited to five years. At the time of writing, the exact requirements were still not fully clear. The draft regulation of the German Ministry of the Environment included requirements for the provision of reactive power and the capacity to support voltage dips (BMU, 2009).

3.7 LOCATION-SPECIFIC DESIGN

Especially in jurisdictions with an extended territory and large differences in the 'quality' (e.g. wind or insolation levels) of locations, we recommend the implementation of location-specific tariffs which differentiate tariffs accordingly. So far, this feature is only in use for wind energy but, in principle, could be transferred to PV. Location-specific tariffs can facilitate a more even distribution of renewable energy plants in a given territory without creating windfall profits for producers at very good locations.

When setting the tariffs, the legislator has to make sure that the profitability of projects at 'good' sites is still higher than at relatively 'bad' sites. Otherwise, the overall efficiency of the FIT scheme might be seriously lowered. Despite the provision of location-specific tariffs, producers should still be given an incentive to search for the best locations.

Location-specific remuneration implies the necessity to evaluate the quality of a location. In the case of wind energy, measurement usually takes place during the first years of operation, as measurements before the start of a project would seriously delay projects. During this period, the remuneration is the same for all producers (a 'flat tariff'), regardless of the quality of the site. The average quality of a site is measured (generally over the first five to ten years), and this determines the tariff level in the later years.

The choice of location-specific tariffs depends to an extent on the population density in a given territory. In densely populated countries, the policy maker will feel the necessity to distribute wind power generation plants evenly in order to avoid social problems such as public acceptance. In this case, location-specific tariffs for wind energy are certainly the best solution, as deployment will not only take place at a few, very windy spots. At the same time, the already installed capacity might influence the legislator's choice for implementation. If the best spots are already taken one can either increase the tariffs in general or implement location-specific tariffs.

The French system of location-specific tariffs for wind energy is relatively easy to understand and adopt. It is based on explicit formulae of the 'profitability index method' (Chabot et al, 2002) (see Section 2.3.2) and established different tariffs according to the full load hours of each individual wind power plant. In France, tariff payment is guaranteed for 15 years. During the first ten years, all producers receive a fixed tariff of €0.082/kWh. During the final five years of operation, the tariff payment depends on the annual, average electricity generation in the first ten years. The power generation is measured in full load hours. In order to get a good average value, the best and the worst year are excluded from the calculation. According to the full load hours at each location, the tariff payment can vary between €0.028 and 0.082/kWh, as shown in Table 3.2.

Table 3.2 *Location-specific remuneration for onshore wind power under the French FIT scheme*

Annual duration of reference operating time	Tariff for the 10 first years (€/kWh)	Tariff for the 5 following years (€/kWh)
2400 hours and less	0.082	0.082
Between 2400 and 2800 hours	0.082	Linear interpolation
2800 hours	0.082	0.068
Between 2800 and 3600 hours	0.082	Linear interpolation
3600 hours	0.082	0.028

Source: Roderick et al, 2007

3.8 Tariff Payment for Limited or Total Electricity Generation

As an alternative to location-specific tariffs, the policy maker can avoid windfall profits for producers at very good locations by limiting the tariff payment to a certain number of produced electricity units. This approach has been followed by the Portuguese legislator in the case of wind energy, small hydro and solar PV. In Portugal, tariff payment is guaranteed for 15 years. However, if the power producer reaches a set maximum amount of electricity beforehand, the FIT payment will stop automatically. The limit has been set at 33,000MWh/MW for wind power plants; 42,500MWh/MW for small hydro; and 21,000MWh/MW for solar PV. In order to implement this design option, the legislator needs to have a very good idea about the total electricity production capacity of installations in a given country. This needs to be evaluated in the first years of the FIT scheme.

This design option might also be useful to avoid fraud. Under the French FIT scheme, this limitation has been implemented for this purpose in the case of solar PV. The tariff payment is guaranteed for a maximum of 1500 hours in service on the French mainland. In the French Overseas Departments and Territories the solar PV tariffs are paid for a maximum of 1800 hours in service. The legislator was afraid that solar PV systems might be combined with a battery, thus increasing the total amount of produced electricity. Instead of controlling each PV installation to prevent illegal combination with batteries, this limit was introduced.

The responsible Ministry assumed that the geographic conditions in France do not allow for longer maximum service hours. However, in the case of the French FIT scheme, this limit blocked important innovations. On the French mainland, for instance, solar PV installations can have more than 1500 hours in service as soon as they are combined with solar tracker devices which orient the solar collector towards the sun. Therefore, the French Ministry is to modify this regulation.

3.9 Tariff Degression

Tariff degression means that FITs are reduced automatically on an annual basis. This reduction, however, only affects new installation. In other words, once a power plant is installed the tariff payment remains constant over a long period of time despite tariff degression. If the legislator decides to amend the FIT legislation periodically, e.g. every four years, tariff degression allows for automatic reduction of the remuneration rate in the meantime without the negative effects of a lengthy political decision-making process. For instance, in 2009 a solar PV plant in a given country might be granted a tariff of €0.3/kWh for the following 20 years. Assuming an annual degression rate of 10 per cent, the tariff payment for installations connected in 2010 will only be of €0.27/kWh. Therefore, tariff degression also stimulated investors to speed up the planning process: the sooner you get connected to the grid, the higher will be the tariff payment for the power plant.

Germany was the first country to implement this design option in order to both anticipate technological learning and provide an incentive for the industry to further improve renewable energy technologies. The cost reduction potential of renewable energy technologies is based on economies of scale and technological innovation. In the last decade, the generation costs for wind and solar power, for instance, dropped by over 50 per cent. In line with the remaining learning potential of each technology, a low or a high degression rate is fixed by the legislator. Relatively mature technologies, such as wind energy, have either a very low degression rate or no degression rate. In Germany, for instance, the tariff is automatically reduced by 1 per cent every year. Technologies whose generation costs are still declining rapidly will need to have a higher degression rate. The degression rate in Germany for solar PV can be up to 10 per cent per year. Table 3.3 shows the full list of German degression rates for the different technologies in 2009.

Table 3.3 *Tariff degression rates under the German FIT scheme*

Renewable energy technology	Annual degression rate (per cent)
Hydropower (more than 5MW)	1
Landfill gas	1.5
Sewage treatment gas	1.5
Mine gas	1.5
Biomass	1
Geothermal	1
Wind power offshore	5 (from 2015 onwards)
Wind power onshore	1 per cent
Solar PV	8–10

Source: Own elaboration based on BGB, 2009

3.10 FLEXIBLE TARIFF DEGRESSION[1]

Flexible tariff degression is a new design option which links the degression rate to the market growth of a given technology (Jacobs and Pfeiffer, 2009). As described above, the cost reduction potential of a certain technology is partly due to economies of scale. If the installed capacity in a given country increased significantly, one can expect production costs to decrease similarly. Besides, the installed capacity and deployment rate for a given technology is a good indicator of whether the tariff level is too high or too low. In the case of unnecessarily high tariffs, producers will have a strong incentive to invest and national objectives might even be exceeded. Thanks to the flexible degression, tariffs will be lowered automatically. In the case of low tariffs, the newly installed capacity will decline or come to a complete standstill. The implementation of the flexible degression feature will automatically lower the annual degression rate and spur investments again.

In order to make this work, the legislator will first have to define a standard degression rate (see previous section). The deviation from this standard degression rate can either be linked to a predefined pathway of market growth (e.g. 25 per cent growth per year) or a given capacity target (e.g. 400MW newly installed capacity per year). If those targets are exceeded the degression will be increased; if they cannot be achieved the degression rate will be reduced. The evaluation of the actual market development requires the establishment of a national registry where all new projects must sign up. The statistics gathered on market growth will provide the basis for setting the annual degression rate.

The volatility of the degression rate might have to be limited by setting maximum deviations from the standard degression at both the top and the bottom end. For instance, if the standard degression rate is 5 per cent, the deviation at the top or bottom is limited to ± 3 per cent, creating an overall 'corridor' of 2–8 per cent.

This design option was implemented in Germany and Spain in 2008. In both cases, the new regulation only affected solar PV, a technology with a large potential for future cost reduction. Accordingly, we recommend considering the implementation of this design option only in rapidly developing markets, e.g. the market for solar PV. For most other technologies, standard tariff degression will be sufficient to control windfall profits. Besides, a sufficiently large national market has to be established in order to estimate the future market growth or capacity target.

In Germany, the maximum deviation from the standard degression rate (see previous section) is only ± 1 per cent. The legislator fixed a predefined pathway of market growth until 2011, expecting the installed capacity to reach 1500MW in 2009, 1070MW in 2010 and 1900MW in 2011. Depending on the actual installed capacity in those years, the degression rate can either be reduced or increased, as shown in Table 3.4.

Table 3.4 *Flexible degression under the German FIT scheme*

Year	Additional capacity in 2009	Deviation from standard degression rate
2009	Less than 1000MW	Minus 1 per cent
	Between 1000MW and 1500MW	No deviation
	More than 1500MW	Plus 1 per cent
2010	Less than 1100MW	Minus 1 per cent
	Between 1100MW and 1700MW	No deviation
	More than 1700MW	Plus 1 per cent
2011	Less than 1200MW	Minus 1 per cent
	Between 1200MW and 1900MW	No deviation
	More than 1900MW	Plus 1 per cent

Source: Jacobs and Pfeiffer, 2009

Previously, the German Greens have suggested linking the degression rate to a 15 per cent standard market growth rate. If actual growth is below or above this expected rate, each percentage point would be translated into an increase or decrease of the degression of 0.1 per cent. For instance, if the national market grows by 20 rather than 15 per cent, i.e. 5 per cent more than expected, the degression rate would increase by 0.5 per cent.

Similarly, the Spanish legislator operates with a predefined pathway of market growth. However, the Spanish mechanism is even more complicated as the degression adjustment occurs every four months and not just once a year.
In order to avoid (further) confusion, we only present the first proposal of the Spanish PV association, as it shows clearly how the degression rate could be linked to market growth (ASIF, 2008). The ASIF proposal envisaged a standard degression rate of 5 per cent in the case of a 20 per cent market growth. The deviations from this standard degression are shown in the Table 3.5.

Table 3.5 *Flexible tariff degression as proposed by the industry association ASIF (Spain)*

Market growth	Degression rate
≤ 5%	2%
10%	3%
15%	4%
20%	5%
25%	6%
30%	7%
35%	9%
≥ 45%	10%

Source: Jacobs and Pfeiffer, 2009

3.11 Increasing Tariffs

Usually, the policy maker can count on reducing tariffs with each amendment, as technology learning and economies of scale will normally lead to production cost reductions (see Section 3.9). In some cases, however, it might be necessary to increase tariff payment after a couple of years. This is especially true for technologies which have already reached a high degree of technological maturity such as wind energy.

In the early years of a FIT scheme, wind energy producers will choose the best national locations for their turbines in order to maximize incomes. In the following amendment rounds the policy maker will have to take into account that the best spots might have already been taken. Less windy locations, however, require higher tariff payments. To a certain extent the need for higher tariffs might be compensated by technological learning as wind turbines can nowadays also operate profitably under less optimal conditions. Nonetheless, the experience in Spain shows that tariffs for wind power had to be increased.

The old FIT scheme of 2004 was based on 2400 full load hours for wind turbines. In the new law of 2007, the responsible ministry based the tariff calculation on 2200 full load hours, believing that the national target can be achieved by using all locations with the respective wind speed. The fixed tariff for wind energy was increased from €0.068 to 0.073/kWh. The same happened in Germany. In 2009 the higher initial tariff was increased from €0.0787 to 0.092/kWh. Tariff increase in both countries was partly due to higher prices for raw materials (mainly steel) on the world market.

To put it in a nutshell: policy makers should not be stubborn when adjusting tariffs. Even though there is certainly a downward trend in generation costs for renewable energies, the legislator will have to observe national and international market conditions, and base its decisions on empirical grounds, not on wishful thinking.

3.12 Inflation-Indexed Tariffs

The general lifetime of renewable energy power plants is in the range of 20–30 years (except hydropower), although efficiencies may dip in later years. Under good FIT legislation, the time for full cost recovery and paying all debts usually takes 15–20 years. As a matter of fact, such long-term investment projects are very sensitive to inflation effects. Therefore, it might be necessary to adjust tariff payments to annual inflation. This is especially important for countries with a high annual inflation rate, as after several years the 'real' remuneration rate will be significantly lower than the nominal rate as fixed by the FIT scheme. Inflation indexation affects both old and new installations.

When you plan to implement the inflation indexation you will need to decide whether you want to fully or just partly protect tariffs against inflation. Full inflation indexation has been implemented in some countries, including Ireland. The Irish legislator fully adjusts the tariff level to the consumer price index on an annual basis. In Spain, the legislator has decided to only partly adjust tariffs to inflation. Until December 2012, the tariffs are adjusted annually to the inflation index minus 25 per cent. From 2013 onwards, the tariffs are adjusted to the national inflation index minus 50 per cent. This leads to minor tariff degression for existing plants. You might also consider indexing tariffs to other economic parameters which have an impact on the cost of power production, such as the price of raw materials (e.g. steel) or labour. This approach has been followed by the French legislator.

If you decide to only partially adjust tariffs or not to adjust them at all you will have to take the estimated future inflation effects into account when calculating the tariffs (see Section 2.3). Since forecasting the inflation effects for the coming 15–20 years is a difficult task, inflation indexation is strongly recommended.

3.13 Additional Tariff Payment for Innovative Features

Many FIT schemes have incorporated small, additional remunerations for a number of technological innovations. For instance, the German legislator grants further bonus tariffs for the use of organic Rankine cycles or thermochemical gasification in biomass plants (€0.02/kWh) and advanced 'hot dry rock' technology in the case of geothermal power plants (€0.04/kWh). Similarly, the French FIT scheme offers an extra tariff of €0.02/kWh for biogas plants with methanization, and up to €0.03/kWh for biomass, biogas or geothermal plants (Klein et al, 2008).

In the following sections, we will take a closer look at bonus payments for:

- repowering of wind power plants; and
- building-integrated PV (BIPV) modules.

3.13.1 Repowering

Repowering refers to the replacement of old wind turbines with new and more efficient machines. Since the locations for harnessing wind power in a given country are limited, the legislator might want to grant an extra incentive for replacing old turbines with new equipment as soon as possible. By repowering, the generated electricity can be significantly increased and the number of turbines can be reduced. Moreover, multi-megawatt turbines turn much more slowly and therefore have a smaller visual impact. Extra payments for repowering are granted in countries which have started to develop wind power projects at an early stage, including Germany, Spain and Denmark. In those countries, the first wind power plants have already been replaced by newer turbines.

In Germany, an extra tariff of €0.005/kWh is granted for repowering. The replaced turbines have to be more than ten years old and located in the same region as the new plants. New plants need to have at least twice the installed capacity of the old plants. In Spain, repowering is remunerated by an extra bonus of €0.007/kWh. The tariff payment is limited to the first two GW of additional capacity and only applies to the replacement of wind power plants constructed before 2001. The tariff payment will remain in place until the end of 2017.

3.13.2 Building-integrated solar panels

Normally, there are three types of tariffs for solar PV depending on the location of the module. Free standing PV systems are located on the ground, modules can be screwed on top of buildings (partially building-integrated) or they can be architecturally integrated. Some countries focus on building-integrated PV (BIPV) modules as they wish to minimize the visual impact. The precise definition of the location of modules varies from country to country.

The Italian FIT differentiates between non-integrated, partially integrated and architecturally integrated solar PV systems. The Decree of 19 February 2007, Article 2.1(a) states that:

> *b1) A non-integrated photovoltaic system is a system with modules located on the ground or with modules located [...] on elements of urban and street furniture, on the external surfaces of the shell of buildings or building structures, whatever their function and intended use;*
> *b2) A partially integrated photovoltaic system is a system of which the modules are positioned [...] on elements of urban and street furniture, on the external surfaces of the shell of buildings or building structures, whatever their function and intended use;*
> *b3) An architecturally-integrated photovoltaic system is a photovoltaic system of which the modules are integrated [...] in elements of urban and street furniture, on the external surfaces of the shell of buildings or building structures, whatever their function and intended use.*

Table 3.6 *Remuneration for solar PV under the Italian FIT scheme*

Size of plant (installed capacity)	Free standing (group b1)	Partially integrated (group b2)	Architecturally integrated (group b3)
Up to 3kW	0.40	0.44	0.49
3–20kW	0.38	0.42	0.46
More than 20kW	0.36	0.40	0.44

Depending on the location of the PV module, the Italian FIT scheme offers differentiated tariffs. Also, the tariffs vary according to the size of the plant. The tariff levels are summarized in Table 3.6.

In Germany, there have also been discussions on granting a bonus tariff payment for thin film PV modules, as they were considered to be an innovation in contrast to standard modules. Finally, however, the legislator decided to be technology-neutral within this one technology area in order to avoid market interventions at an early stage of technological development. This avoided 'picking a winner' and directing technological development in a certain direction without sufficient information about the future potential of each of these technological options.

3.14 EXCLUSION FOR ENERGY-INTENSIVE INDUSTRIES

An increase in the electricity price, even though it is minor, can have severe negative effects on some consumer groups. This might be especially the case for energy-intensive industries, where sometimes up to 80 per cent of total production costs are energy related. Since these companies are often subject to international competition, exemption from the general financing scheme might have to be incorporated in the FIT legislation. It has to be noted that environmental non-governmental organizations (NGOs) have often criticized this exemption as it reduces the necessity for the electricity-intensive industry to lower consumption. Besides, these consumer groups are often able to profit from preferential electricity tariffs.

Exemptions for such industries have been implemented in a number of European countries, including Austria, Denmark, Germany and the Netherlands. Electricity-intensive industries and companies can be identified by measuring the total amount of electricity consumed, or the ratio of the annual costs of electricity consumption to other parameters, e.g. the revenues, the total costs or the gross value added (Klein et al, 2008).

From 2004, German companies with a total electricity consumption of more than 100GWh and total electricity costs exceeding 20 per cent of the gross added value were partly exempt from supporting renewable electricity under the FIT. The increase of the electricity price due to the FIT payment is limited to €0.0005/kWh. In the first years, only 40 companies profited from this exemption. In 2006, the criteria were loosened (10GWh consumption and 15 per cent of the gross added value) and almost 400 companies were exempt. Due to the exemption of these industries, the additional cost for all other electricity consumers increased by 17 per cent in 2007 (Wenzel and Nitsch, 2008).

References

AEE (2008) *Anuario del Sector: Análisis y Datos*, Asociación Empresarial Eólica, Madrid

ASIF (2008) *La Tarifa Fotovoltaica Flexible Reduciría la Retribución Entre un 2 Per Cent y un 10 Per Cent Cada Año*, comunicado de prensa, Madrid, 13 March

BGB (2008) 'Gesetz zur Neuregelung des Rechts der Erneuerbaren Energien im Strombereich und zur Änderung damit Zusammenhängender Vorschriften', *Bundesgesetzblatt*, vol 1, no 49, p2074

BMU (2009) *Verordnung zu Systemdienstleistungen durch Windenergieanlagen (Systemdienstleistungsverordnung SDLWindV)*, Entwurf, BMU – KI III 4, Stand: 2. März 2009

Chabot, B., Kellet, P. and Saulnier, B. (2002) *Defining Advanced Wind Energy Tariffs Systems to Specific Locations and Applications: Lessons from the French Tariff System and Examples*, paper for the Global Wind Power Conference, April 2002, Paris

Couture, T. (2009) *An Analysis of FIT Policy Design Options for Renewable Energy Sources*, Masters Thesis, University of Moncton, available from author

Jacobs (2008) *Analyse des Spanischen Fördermodells für Erneuerbare Energien unter Besonderer Berücksichtigung der Windenergie, Wissenschaft und Praxis*, Studie im Auftrag des Bundesverbands WindEnergie e. V., Berlin

Jacobs, D. and Pfeiffer, C. (2009) 'Combining tariff payment and market growth', *PV Magazine*, May, p220–24

Klein, A., Held, A., Ragwitz, M., Resch, G. and Faber, T. (2008) *Evaluation of Different FIT Design Options*, best practice paper for the International Feed-in Cooperation, www.worldfuturecouncil.org/fileadmin/user_upload/Miguel/best_practice_paper_final.pdf

Roderick, P., Jacobs, D., Gleason, J. and Bristow, D. (2007) *Policy Action on Climate Toolkit*, www.onlinepact.org/, accessed 23 May 2009

Wenzel, B. and Nitsch, J. (2008) *Ausbau erneuerbarer Energien im Strombereich, EEG-Vergütungen, Differenzkosten und – Umlagen sowie ausgewählte Nutzeneffekte bis 2030*, Teltow/Stuttgart, Dezember 2008

4

Bad FIT Design

Many of the pitfalls of FIT legislation have already been implicitly described in the previous chapters, where we have evaluated best practices for basic and advanced FIT design options. In other words, if you stick to the design options we have recommended in the previous section you should be able to avoid the major drawbacks, including high costs for grid connection, inflated costs for the final consumer, and unnecessary administrative barriers.

Nonetheless, the negative experiences of some countries have urged us to include a chapter on bad FIT design. This will elaborate on a few design options which have proven to be especially counterproductive when it comes to a fast uptake of renewable energies, as well as longer-term sustainability of the industry. This will also help readers understand why we have suggested certain design criteria in the previous chapters. Furthermore, we are going to show you who is usually behind the push towards 'bad FIT design'.

4.1 Low Tariff Level

Remember: a low tariff is almost like no tariff. The worst that can happen is that after a lengthy campaigning and political decision-making process, you end up with a FIT scheme but the tariffs are too low. With a very low tariff, few if any will feel inclined to invest in renewable energies because projects will not be profitable. You might still have a few ecologists who would have invested even without any tariff payment at all, but you will not see a massive uptake of renewable energies in your country. Argentina has had this kind of negative experience with low tariffs. In December 2006, after a series of failed attempts by several senators, the congress passed a FIT scheme. Due to political opposition, however, the tariff level was set very low. Wind power was granted a premium tariff payment of €0.37/kWh – far too low for any serious investment. Predictably, the installed wind capacity has remained 'stable' at 30MW for years now.

Who would argue for low tariffs? Well, probably the same people who were opposed to renewable energy support and FITs in the first place. In the 1990s, when wind power started to grow considerably in Germany and the utilities started to realize that wind power could ultimately be a threat to their dominance of the energy market, the large utilities started to argue that the tariff level was too high. In this case, it will be crucial to have a least a small number of dedicated people in the parliament or the ministries who will have the final word when it comes to fixing the tariff. Sometimes, governmental actors will also try to push for low tariffs because they are afraid of high costs for the final consumer. Under those circumstances, a clear and transparent methodology for tariff level calculation will help to establish fair and sufficient tariffs (see Section 2.3).

If you should enter the race for a green economy with such a false start don't be overly disappointed. There are many examples of countries where the tariff level has been too low at the start. If you manage to keep up the pressure and inform the government, politicians and public about the bad performance of the national support scheme, an improvement of the tariff level in the next amendment round will be very likely. Therefore it is crucial to establish periodical progress reports in the first FIT legislation (see Section 2.13).

4.2 Unnecessarily High Tariff Level

Unnecessarily high tariffs can seriously undermine the efficiency of your support instrument as you will pay more than necessary. Again, a clear and transparent tariff calculation methodology (see Section 2.3) and an analysis of other FIT countries can help to avoid this pitfall. Extremely high tariffs might eventually affect the overall stability of your national FIT scheme, as political and public support are at risk due to the unnecessary costs.

Unnecessarily high tariffs in one country can also have negative effects on the renewable energy support in other countries. In a globalized economy, producers of renewable energy equipment will try to sell their products to countries which have the best possible conditions. If the rates of return in your country are significantly higher than in other countries your national markets might absorb too large a percentage of products on the world market. Consequently, other countries might have difficulties in achieving their national targets.

Spain experienced these negative effects in the last few years. Even though the tariffs for solar PV were similar to the German tariff level it was much higher in 'real terms', as the sunshine intensity in Spain is about 60 per cent higher than in Germany.[1] Even with the tariff reduction in 2008 it was estimated that the Spanish FIT was far more attractive than the German tariff. Due to the comparatively high tariff for solar PV and very good climatic conditions in Spain, the market grew by more than 500 per cent in 2006 (ASIF, 2007). In 2007, more than 2.5GW of new solar capacity were installed, almost half of the worldwide installed capacity (EPIA, 2009)! In the light of this unsustainable growth, the renewable energy industry associations had few arguments when they tried to convince the national legislator to remove the capacity limit in 2008. It has to be noted, however, that the vast amount of installed capacity in Spain was due to a combination of two 'bad' design options: extremely high tariffs and a capacity cap (see Section 4.8).

Who will argue for an extremely high tariff? It seems logical that the renewables industry will argue for high or even extremely high tariffs: the higher the tariffs, the higher the profitability margin. However, people with this attitude should be warned that short-term profitability can sometimes endanger long-term objectives. The main objective of renewable energy industry associations should be the transformation of the entire power sector, and not unsustainable internal rates of return.

4.3 FLAT-RATE TARIFF

A flat-rate tariff, which gives the same fixed price for all renewables, contradicts the basic idea of a FIT. When you choose to support renewable energies with a FIT scheme you choose a price-based support mechanism which allows you to establish technology-specific support. Technology-specific support is one of the main advantages of FITs over quantity-based support instruments because it allows you to reduce windfall profits. Almost all studies evaluating the effectiveness and efficiency of support instruments argue that technology-specific support is crucial if you want to minimize costs (Ragwitz et al, 2007). Why would you support this heterogeneous group of renewable energy technologies under a flat-rate tariff if you know that the generation costs of each technology are completely different? Only because you can group them together under the generic term 'renewable energies'? No country would base its agricultural policy on a flat-rate payment per kilojoule just because all agricultural products fall into the category of 'nutrition'.

The US PURPA Act of 1978 was technology-neutral. Equally, several European countries, for example Estonia, operate FITs with a single flat-rate tariff, but without much success. If you make the effort to establish a FIT scheme it does not take much to establish different tariffs for different technologies. The lack of technology-specific tariffs is sometimes due to the wrong tariff calculation methodology (see Section 2.3). In Hungary, for instance, wind and solar PV receive the same tariff payment even though the generation costs are not remotely in the same range. Conversely, establishing a tariff calculation methodology based on the technology-specific generation costs and sufficient profitability will automatically lead to technology-specific support.

If you hear people arguing for flat-rate tariffs, check if they are producers of relatively cost-efficient renewable energy technologies. It certainly makes sense for them to argue for a flat-rate tariff as the tariff will probably be higher than their average generation costs. A general argument for flat-rate tariffs is the intention to spur competition between renewable energy technologies. Hearing this argument we will have to ask whether we want to establish a 'kindergarten' for renewable energy technologies where they compete against each other, or whether we are supporting renewable energies in order to make them competitive with conventional power production technologies. Coming back to the previous comparison we can also ask: will you establish competition between the producers of butter and bread by granting them the same level of support?

A large portfolio of renewable energy technologies will be crucial in the future. We cannot afford to support only the most cost-effective solutions today if we know that we will probably need all technologies in the future. FITs enable policy makers to develop all renewable energy technologies at once even though they are at different levels of technological development.

4.4 Maximum and Minimum Tariffs

Minimum tariffs have the great advantage that lengthy negotiations about the power purchase agreement can be avoided. This significantly reduces transaction costs. If you set a maximum tariff, as reported from Kenya, this advantage is lost as grid operators and producers have to agree on a tariff payment. Therefore, it is important to stress that the established tariffs are minimum tariffs. This is also important for technologies which have almost reached competitiveness with conventional power generation technologies. Otherwise, some people might argue that as soon as the market price for electricity surpasses that tariff level no more tariff payment shall be granted.

4.5 Exemptions from Purchase Obligation

As described in Section 2.8, the purchase obligation is one of the most important elements of FIT schemes. In combination with the fixed tariff payment, the purchase obligation offers very high investment security, which can be seriously offset by exemptions from this obligation. In Estonia and Slovakia, for instance, grid operators are only obliged to purchase renewable electricity up to the level of their transmission and distribution losses. In Estonia, this is due to the fact that not all network operators have a licence to sell electricity. Consequently, they can only purchase the electricity necessary to compensate for their network losses, which are generally low in times of low consumption. This causes great insecurity for green power producers (Klein et al, 2008).

The Kenyan FIT states that 'Power Producers and grid system operators may agree by contract to digress from the priority of purchases, if the plant can thus be better integrated into the grid system' (Kenyan Ministry of Energy, 2008). Negative effects from this exemption can only be avoided if this provision does not come into force if one of the parties does not explicitly favour it. Otherwise, the grid operator can always refer to this exemption clause and thus undermine investment security.

4.6 Bad Financing Mechanisms

In the past, many problems with FITs have occurred when financing was not exclusively guaranteed by a top-up on the electricity bill of the final consumer (see Section 2.7). Alternatively, governments have tried to finance FITs through taxes, the general state budget or a fund scheme. All three approaches have proven to be problematic as governmental money is included. Because of this, FITs easily become the subject of political debates when the government changes or the national economy goes through difficult times. As shown in Section 5.3, the inclusion of governmental money might, however, be necessary for FITs in developing countries.

The Spanish FIT scheme is largely financed by a top-up on the final consumer's electricity bill. However, a small part of the additional costs is paid by the general national budget as electricity prices in Spain are still regulated and do not fully mirror supply and demand. The annual deficit of the Spanish electricity system is covered by the general budget. The deficit has increased to more than €10 billion since 2000 (Libertad Digital, 2009). This is one of the reasons why the Spanish government decided to keep the cap for solar PV. Recently, South Korea has decided to phase out the FIT scheme in 2012 and replace it by a renewable portfolio standard mechanism (see Section 9.1). The government argued that the costs of the FIT scheme were too high. Not surprisingly, the FIT scheme in South Korea is financed via taxpayers' money (Sollmann, 2009).

4.7 Bad Tariff Calculation Methodologies

In contrast to the successful 'generation cost' approach (see Section 2.3), FITs are sometimes calculated based on other indicators, namely average electricity prices and avoided costs. These alternative tariff calculation methodologies were mostly applied in early FIT schemes in the 1990s. However, some legislators are still designing FITs around these concepts.

These tariff calculation methodologies have proven to be less successful. The reason for this is relatively simple. As stated above, a policy maker designing a FIT scheme is looking for a balance between investment security and sufficient returns on investment for the producer on the one hand, and minimum additional costs for the final consumer on the other. The most straightforward approach is to take the technology-specific generation costs into account. In the case of the alternative tariff calculation methodologies, the tariff level will only coincidentally match the costs for the producer. There is a high risk, however, that the tariff level will either be too low or too high. If the tariff is too low, no investment will be triggered and the FIT scheme will be ineffective. If the tariff is too high, investments will be made but the costs for the final consumer will be unnecessary high due to windfall profits. Therefore we cannot recommend the approaches described below.

4.7.1 Avoided costs

In the case of the 'avoided cost' approach, two different methodologies have to be distinguished. First, some policy makers refer to the 'avoided costs' from conventionally produced electricity, i.e. the cost that would have occurred if the renewable electricity had been produced by other, conventional generation technologies such as gas-fired or nuclear power plants.

Accordingly, the French FIT calculations are based on the long-term avoided costs. However, renewable electricity producers receive an additional top-up payment if additional public service objectives are met. The additional objectives are defined as 'energy independence and security of supply, air quality and greenhouse gas emissions, the optimal management of national resources, the management of future technologies and rational use of energy' (*Journal Officiel de la République Française*, 2000). Due to this top-up, most technologies receive a sufficiently high tariff. However, the tariff calculation methodology is unnecessarily complicated and often leads to confrontations between various governmental actors who operate with other tariff calculation methodologies (see Section 2.3.2).

Besides the above-mentioned problems, this approach often does not allow for sufficiently high tariffs for less mature technologies. While the tariffs under this approach might be high enough – or even too high – for relatively cost-effective technologies such as wind energy, technologies like solar PV will probably not benefit.

In addition, some countries refer to the 'avoided external costs', i.e. the negative environmental and social impact of conventional energy production that can be avoided through the deployment of renewable energies. This approach has partly been followed in Portugal. More precisely, the Portuguese FIT scheme takes into account the avoided capital investment, and the avoided costs in electricity production (fuel, operation and maintenance), network losses, and to the environment (IEA, 2008).

To base the tariff calculation on the 'avoided costs' also carries the risk of the varying interpretation of this term. While the 'avoided costs' of conventional energy generation in a given market can still be calculated relatively objectively, the estimate of the 'avoided external costs' is based on a large number of assumptions. Tariff levels might therefore be subject to the government's position on climate change and environmental protection.

4.7.2 Electricity prices

Some countries have also chosen to link the tariff payment to the electricity price. In the past, this approach has been used by both Germany, beginning in 1990, and in Spain. The tariff payment was defined as a percentage of the electricity price for final consumers. In the first years, solar PV and wind power received 90 per cent, hydropower, sewage treatment gas, landfill gas and biomass up to 80 per cent. Only wind power and hydropower could considerably increase the installed capacity under this approach. For all other technologies the tariff payment was too low. Germany abandoned this methodology in 2000, as the electricity price started to fall significantly at the end of the 1990s as a short-term result of the European liberalization of energy markets.

In Spain, FITs were linked to the average electricity price until 2007. The average electricity price was fixed on an annual basis and published in a Royal Decree. In 2004, solar PV received up to 575 per cent, onshore wind, hydropower, geothermal and biomass up to 90 per cent. In contrast to the German case, the Spanish legislator decided to replace this approach by a fixed tariff scheme because of soaring electricity prices. In the years 2006 and 2007, the average electricity price increased significantly due to resource scarcity on the world markets. In sum, the biggest problem with this approach is the fluctuation of the electricity price over time. Therefore, we do not recommend this tariff calculation methodology.

4.8 Capacity Caps

Capacity caps limit the total amount of newly installed capacity and consequently hinder the creation of mass markets. FITs are not only granting financial assistance to producers of renewable electricity, but have proven to be the most successful instrument for spurring rapid technological development. One of the major achievements of FITs is that they contribute to driving down generation costs for all renewable energy technologies, making them affordable to a larger number of people and countries in the world, and competitive with conventional energy generation.

This development has been possible because FITs triggered mass production and thus created economies of scale. Classical R&D funding and limited investment subsidies, which were the major support instruments in the 1970s and 1980s, did not spur sufficient investment. Renewable energies have now managed to overcome the stigma of only being applicable to niche markets and have entered into industrial production. The 100 per cent renewable energy scenario is no longer a dream (Schreyer et al, 2008; Droege, 2009).

Capacity caps also have a negative effect on the development of national markets. The experience in many countries has shown that shortly before reaching the cap, the market heats up as all producers race to get connected to the grid. This may lead to exponential market growth within a very short period of time, as was the case for the Spanish solar PV market in recent years. Once the cap is reached the market usually collapses as no more capacity can be installed for a long time. These 'stop and go' cycles prohibit the creation of a national market which requires sustainable and predictable market growth, and stable supply chains.

Total capacity caps are generally considered to be the easiest way to limit the additional costs for the consumer. By contrast, and with regard to the above-mentioned disruptive effects, we recommend other design options, especially tariff degression or even flexible tariff degression. Together with differentiated tariffs, these design options enable the legislator to control market growth and consequently costs. A total capacity cap should only be the last resort, e.g. for developing countries with very limited financial means (see Section 5.1).

Besides policy makers who fear high additional costs, you will probably find producers of grey electricity in the group of people that argues for capacity caps. Naturally, they want to limit the share of green electricity in order to protect their market shares. Once again, do not give up the fight just because your first national FIT scheme includes some sort of cap. Many successful FIT schemes have started operation with capacity caps, including Germany, France and Spain. In contrast to the 1990s, today it will be much easier to argue against capacity caps, as the need for massive renewable energy deployment cannot be dismissed in the light of global climate change.

4.9 Legal Status

Finally, we want to underline the importance of legal status for any type of FIT legislation. Activists and FIT advocates should fight for FITs being established by law and not just by a ministerial order or a vaguely articulated policy. Even though the political decision-making process usually takes longer, it helps to protect FITs from attacks from opponents. For example, renewable electricity producers under the Spanish FIT scheme have had bad experiences with their FITs not having the legal status of a law. In 2006, legislation on urgent measures in the energy sector was passed, which lowered the remuneration for renewable electricity producers by decoupling the tariff level from the average electricity price (*Boletín Oficial del Estado*, 2006). This intervention affected both existing and new plants, thus undermining the stability and predictability of the Spanish FIT scheme. Therefore, the national renewable energy associations are constantly urging the Spanish government to put all related legislation into a single law.

REFERENCES

ASIF (2007) *El Sector de la Solar Fotovoltaica Rechaza el Nuevo Decreto sobre esta Energía Renovable*, comunicado de prensa, 3 October, Madrid

Boletín Oficial del Estado (2006) Real Decreto-Ley 7/2006, de 23 de Junio, por el que se Adoptan Medidas Urgentes en el Sector Energético, http://vlex.com/vid/adoptan-medidas-urgentes-energetico-20769872

Droege, P. (ed.) (2009) *100 Per Cent Renewable: Energy Autonomy in Action*, Earthscan, London

EPIA (2009) 2008: 'An exceptional year for the photovoltaic market', Press Release, 24 March, Brussels

IEA (2008) *FITs, Making CHP and DHC Viable – Portugal Case Study*, IEA/OECD, www.iea.org/g8/chp/docs/portugal.pdf

Journal Officiel de la République Française (2000) 'Loi no 2000-108 du 10 février 2000 relative à la modernisation et au développement du service public de l'électricité', *Journal Officiel de la République Française*, vol 35, 11 February, p2143

Kenyan Ministry of Energy (2008) *Feed-in-Tariff Policy on Wind, Biomass and Small-Hydro Resource Generated Electricity*, Ministry of Energy, Kenya

Klein, A., Held, A., Ragwitz, M., Resch, G., and Faber, T. (2008) *Evaluation of Different FIT Design Options*, best practice paper for the International Feed-in Cooperation, www.worldfuturecouncil.org/fileadmin/user_upload/Miguel/best_practice_paper_final.pdf

Libertad Digital (2009) *El Gobierno Pone Fin al Déficit Eléctrico y Congela la Tarifa a las Rentas Bajas*, 30 April, www.libertaddigital.com/economia/el-gobierno-pone-fin-al-deficit-electrico-y-congela-la-factura-a-las-rentas-bajas-1276357914/

Ragwitz, M., Held, A., Resch, G., Faber T., Haas, R., Huber, C., Coenraads R., Voogt, M., Reece, G., Morthorst, P. E., Jensen, S. G., Konstantinaviciute, I. and Heyder B. (2007) *Assessment and Optimisation of Renewable Energy Support Schemes in the European Electricity Market*, Final Report, OPTRES, Karlsruhe

Schreyer, M., Mez, L. and Jacobs, D. (2008) *ERENE – Eine europäische Gemeinschaft für Erneuerbare Energien*, Band 3 der Reihe Europa, Heinrich-Böll-Stiftung, www.greens-efa.org/cms/default/dokbin/239/239844.pdf

Sollmann, D. (2009) 'Ende für Einspeisetarife, Südkoreas "grüne Wachstumsstrategie" streicht die Einspeisetarife und fördert die Atomkraft', *Photon*, April, pp26–27

5

FIT Design Options for Emerging Economies

Providing access to essential energy services (such as food, light, comfort, communications and cold beer) is one of the driving forces of socio-economic development and poverty reduction. Worldwide, nearly 2.4 billion people use traditional biomass fuels for cooking and heating, and 1.6 billion do not have access to electricity. Their energy needs have to be fulfilled by the use of very basic resources, such as fuel wood, charcoal, agricultural waste and animal dung. These low-level forms of energy supply have a significant negative impact on human health. Indoor air pollution, resulting from burning fuels in badly ventilated or enclosed spaces, contributes to the death of about 1.6 million people every year according to the World Health Organization (WHO, 2005).

In almost all regions with low electrification rates there is, perhaps surprisingly, a large potential for renewable energy. In these countries and regions, FITs can play an important role as they are simple in design and easily adaptable to all sorts of energy market frameworks. Most other support mechanisms require very high levels of regulation and successful steps towards the liberalization of energy markets. Tradable certificates, for instance, work best in countries with a large number of power producers in order to assure liquidity of the certificate market (see Chapter 9). In the case of monopolies or oligopolies, actors are tempted to use their market power to manipulate the certificate price. Some types of FITs, such as premium FITs (see Section 3.1), need well-functioning spot and intra-day markets. Once again, liquidity of those markets is essential to avoid volatility of prices and to guarantee relatively stable incomes for producers.

In many cases, FITs fit best into the existing organizational structure of power markets, as developing countries normally have monopolized energy markets or are in the process of transition towards more liberalized markets. In fact, FITs function similarly to the concept of cost recovery used in the rate regulation of utilities in monopolized markets. In fully monopolized or oligopolistic markets, the combination of a purchase obligation and a fixed tariff payment sufficiently protects new actors in the power production business from abuse of market power by established actors. However, readers should remember that the dominant players can still use their market power when it comes to grid connection procedures and administrative barriers (see Sections 2.9 and 2.11). These issues can only be tackled by the regulator. As an additional benefit, FITs can help to break up monopolies and establish new actors. By establishing regulation for grid connection and electricity purchases, FITs provide a legal basis for market access for independent power producers.

To help emerging economies capture these benefits, this chapter discusses FIT design especially for developing countries and emerging economies. A growing number of developing countries and emerging economies has recently implemented or planned to implement FITs, including Argentina, Brazil, China, Ghana, Israel, Kenya, Nigeria, Pakistan and South Africa. The operation of FITs in developing countries and emerging economies follows the same rules as in any other country (see Chapter 2). However, certain additional measures to control the additional costs for the final consumer might have to be considered. This can be achieved through special financing mechanisms (FIT fund), a capacity cap or combination with the Clean Development Mechanism (CDM). Besides, FITs in emerging economies or developing countries will probably have to be inflation-indexed (see Section 3.12) as average inflation tends to be higher than in many industrialized countries.

5.1 Capacity Caps for Developing Countries

FITs have spurred massive investment in renewable energy technologies and have helped to push renewables from niche markets into large-scale industrial production. The full benefits of economies of scale and the resulting technology development can only be captured when no limits are imposed on the growth of renewable energies. We usually recommend no limitation on renewable energy deployment and consider caps in FITs to be bad FIT design (see Section 4.8). However, exceptions can be made in developing countries, since there is a direct link between electricity costs for the final consumer and the amount of electricity that is supported under the FIT. If the share of renewable electricity is limited under the FIT scheme, the additional costs for the final consumer can be easily controlled. In developing countries, increases in the electricity price generally have a strong impact on the final consumer, as people already spend a large share of their income to cover daily energy needs. Therefore, increases in the electricity price are politically sensitive issues.

Developing countries can contribute to the global effort against climate change by constantly increasing the share of renewable energies. However, due to financial restrictions, technology development is not their main objective. Industrialized countries should be committed to this aim. This approach allows for some kind of burden sharing between industrialized countries who have the financial capacity to drive down costs of renewable energy technologies, and developing countries which can profit from cost reduction to meet their increasing energy demand in a sustainable manner. Besides, energy markets in developing countries are often monopolized or have just started the transition towards liberalization. Therefore, the future power capacity development is often subject to national planning. In order to fit renewable energies into these concepts, capacity caps might be necessary.

Capacity caps have to be implemented carefully. Generally, the limits are set for each technology, e.g. 400MW for wind power, 200MW for biomass and 100MW for solar PV. Once the limit is reached, no more tariff payment is granted for new installations. In order to effectively control costs, technologies which are relatively cost-effective should be granted higher share than technologies with relatively high costs. In the case of technology-specific limitations, it might be wise also to limit the size of each plant. If, for instance, you should only grant tariff payment for solar PV up to a total installed capacity of 50MW you will probably not want to see two large solar PV plants of 25MW. The objective is that a large number of producers and users can profit from the FIT scheme. Therefore, you could limit the maximum capacity of each plant eligible under the FIT scheme. This approach has been followed under the Kenyan FIT scheme which grants tariff payment for wind energy up to a total capacity of 150MW. At the same time, a wind park, i.e. one wind power installation, cannot be larger than 50MW. By these means, at least three wind power developers will profit from the Kenyan legislation.

When you decide to implement a capacity cap, the FIT legislation should include a provision for revising this cap before it is reached. This way, you can prevent stop-and-go cycles in the production of equipment. The legislation should state that if, for instance, 75 per cent of the cap or target is reached, the ministry or organization in charge has to evaluate the FIT scheme, including tariff level, capacity caps and targets. Based on this assessment, the legislator should weigh up whether an increase in the capacity cap or target is feasible (for more information on these issues see Section 2.13). To put it in a nutshell: don't wait until you reach the cap. The political decision-making process for new targets takes some time and this has to be anticipated.

In order to create a stable national renewable energy economy, stop-and-go cycles in the production of equipment can also be minimized when the overall cap for one technology is divided into various segments over time. For instance, if the FIT scheme foresees a total installed wind capacity of 300MW by 2012, the legislator can decide that every year 100MW should be issued following the 'first-come, first-served' system. This approach was chosen by the Spanish legislator with respect to the PV market, on a quarterly basis. However, this approach is rather complicated and causes high administrative costs since it requires producers to apply for tariff payment in advance and forces the legislator to handle all applications.

5.2 Application Process

When you implement a cap, not all producers might have the chance to become eligible under the FIT scheme, so it is important to introduce a fair and transparent application process. You will have to set up a registry which is usually within the responsibility of the regulator. Depending on the telecommunication infrastructure in your country or region, this registry can also be accessible online. Applicants have to be considered according to the 'first-come, first-served' approach in order to avoid corruption.

5.3 FIT Fund

Another possibility to reduce additional costs for final consumers is to establish a national fund for FIT payment. The sources of the fund can come both from the national budget and from international donors (see Figure 5.1). For the latter, payments to the fund can facilitate the promotion of hundreds of small-scale renewable energy projects. Previously, the promotion of renewable energy projects through international donors was often impeded by the nature of such projects, e.g. small size projects with high transaction costs.

The fund option, however, includes several risks. One of the main success factors for FITs in many countries was the fact that tariff payment is independent from governmental financing. It simply established payment rules among a number of private actors. Due to the fact that the financing mechanism usually distributes all costs among all final electricity consumers, the system is relatively stable in the event of political changes. As soon as governmental money is included in the financing scheme, there will be a stronger 'incentive' for governments to reduce the support of renewable energies, especially in times of economic downturn.

Additionally, the fund has to set aside large reserves as tariff payment has to be guaranteed over a long period of time of up to 20 years. Due to this, the

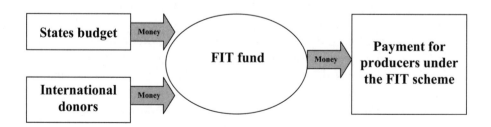

Figure 5.1 *Feed-in tariff fund for developing countries*

Source: Jacobs, 2009

accumulated costs for one renewable energy project might seem rather high and thus attract lower public and political support. Norway planned to finance a FIT scheme by the means of a national fund but finally did not implement this support scheme for that reason (Reiche and Jacobs, 2009). Therefore, financing FITs via a fund should be the option of last resort.

5.4 Cost Distribution Between Final Electricity Consumers

In case you should opt for the normal financing mechanisms of FITs, i.e. the distribution of costs among all electricity consumers, certain variations from the standard procedure in industrialized countries might have to be considered for emerging economies. First, passing on the costs to the electricity bill of each final consumer and even disclosing the additional costs due to the FIT scheme might lead to high administrative complexity and transaction costs. Instead, the distribution system operator can simply add the additional costs to the general pass-through costs of the system which are usually established for independent power production. If this is the case, only the costs exceeding the avoided cost of conventional energy production should be added to the pass-through costs. In order to do so, the avoided costs have to be calculated on an annual basis. This approach was chosen under the Kenyan and South African FIT schemes.

5.5 Combining FITs and the Clean Development Mechanism

The Clean Development Mechanism (CDM) is one of the three flexibility mechanisms for greenhouse gas reductions under the Kyoto Protocol, along with Joint Implementation (JI) and the International Emissions Trading Scheme. It gives industrialized nations, so-called Annex I countries, the opportunity to fulfil their reduction targets through projects in developing countries (Annex II countries). The avoided greenhouse gases are certified and these certificates can hence be traded on the international carbon market in the form of Certified Emission Reduction (CER) units. In past years, renewable energy projects in developing countries have increasingly been co-financed through the CDM. The income generated through the trade of certificates can cover parts of the overall project costs.

However, two major barriers have been identified. On the one hand, investments might be hampered by the relatively high costs and small size of renewable energy projects in comparison to other typical CDM projects, such as energy efficiency. Investors select projects according to the total amount of greenhouse gas emissions that can be avoided and the implied transaction costs. Therefore, large-scale projects with huge emission reduction potential have a competitive advantage over small-scale renewable energy projects (Schröder, 2009). On the other hand, the requirement for 'additionality' can lead to problems when combining the CDM with other support mechanisms for renewable energies, such as FITs.

The additionality criterion says that under the CDM, only those projects that would not have been realized without the additional financial income generated through the trading of certificates on the international carbon market, are eligible. The reason for additionality is to avoid 'free riders'. The Kyoto Protocol was designed to reduce greenhouse gas emissions. If an industrialized country has the chance to reduce its emissions in a developing country this should only be possible if the project would not have been put in place anyway.

Despite the good intentions, in the past the additionality criteria led to perverse effects (Bode and Michaelowa, 2003). In the worst case, policy makers in developing countries might have avoided the implementation of successful policies for renewable energies in order to remain eligible for projects under the CDM. In the light of these adverse effects the CDM Executive Board clarified that national support policies which have been implemented after November 2001 are not included in the baseline calculations (UNFCCC, 2005). Therefore, it is now possible to combine CDM and FITs without any risks.

When calculating the tariff payment under the FIT scheme, the legislator has to decide whether the potential incomes from carbon trade under the CDM will be taken into consideration. Theoretically, one could argue that the additional income can be subtracted from the tariff payment and so the legislator can offer a lower tariff. But when combining FITs and the CDM, problems may arise which are due to the volatility of the certificate price on the international carbon market. As stated above, FITs have proven to be successful because they offer investment security due to fixed tariff payment. If one income component of renewable energy projects is volatile due to supply and demand on the international carbon market, investment security will be reduced.

Therefore, the South African regulator decided not to include carbon revenues from CDM into the tariff calculation, as the Kyoto Protocol will expire in 2012 and there is great uncertainty about the international climate protection regime in the post-Kyoto era. We recommend excluding potential revenues from CDM when calculating the tariffs. The FIT scheme should be 'self-sufficient', i.e. guarantee enough revenues for the producers in order to operate renewable energy units in a profitable manner.

5.6 FITs for Mini-Grids

Hundreds of millions of people have no access to electricity, and most of them reside in areas where no electricity network will be available any time soon. In order to power the local poor, off-grid solutions have to be developed. Instead of extending the national electricity grid to isolated and rural areas at very high costs, mini-grids can provide high-quality indigenous electricity to villages.

Initially, FITs were designed to support grid-connected renewable power generation in industrialized countries. Usually, those countries relied on an expansive electricity grid to tap into renewable electricity potential. However, many countries with a less developed grid infrastructure have a large potential for renewable energies. Therefore, the Joint Research Centre (JRC) of the European Commission has tried to modify FITs so that they are applicable to mini-grids (Moner-Girona, 2008).

Mini-grids refer to interconnected, small-scale, modular electricity networks that rely on a small, local and often isolated distribution system (see Figure 5.2). Typically, a small village or a group of houses is interconnected by such a mini-grid. Local mini-grids can later be extended by linking several village mini-grids in a given area. Electricity can be provided by a large variety of technologies. In most cases, wind turbines and solar panels are backed up by small diesel generators or batteries. Ideally, biomass and small hydro can provide this backup power as well. Such hybrid systems can provide reliable electricity at community level (for more on these benefits, see Chapter 7).

By definition, costs that occur in an isolated mini-grid cannot be shared amongst all electricity consumers in the country. Therefore, the financing mechanism of a FIT scheme (see Section 2.7) has to be adjusted to meet the needs of the actors that participate in such a system. In order to understand the constellation of possible stakeholders, first of all we have to ask whether the mini-grid allows for

Figure 5.2 *Mini-grid powered with renewable energy sources*

Source: Solar Technology AG, with permission from Alliance for Rural Electrification (ARE, 2009)

independent power generation, similar to liberalized electricity markets, or whether one actor has a local or regional monopoly on power generation.

In the case of a 'liberalized' approach, the independent power producer (IPP) is allowed to generate electricity and feed it into the grid. Legally, independent power production is based on long-term contracts, so-called power purchase agreements (PPAs). By now, more than 25 developing countries have set up frameworks for IPPs. The major difference regarding large utilities is that they are not engaged in electricity transmission or distribution activities, but are only active in the generation business.

In order to facilitate renewable electricity generation through IPPs the national regulator or responsible ministry would first set an adequate FIT (see Section 2.3). The IPP can connect to the local mini-grid which is managed by the local distribution system operator (DSO). One part of the IPP's income will come from the DSO who receives the payment from the final consumer. In many cases, however, the local DSO sells electricity to the final consumer at prices below the average generation costs. Therefore, the remaining difference to the FIT will have to come from the national level, e.g. the governmental electricity authority or regulator. If necessary, part of the governmental payment might be facilitated though international donors. Together, both sources of income should be sufficient to guarantee profitability for renewable electricity projects from IPPs (see Figure 5.3).

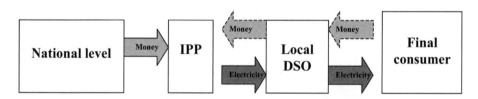

Figure 5.3 *Feed-in tariff for mini-grids (IPP)*

Source: Jacobs, 2009

Analogously, household electricity consumers can produce part of their electricity needs through small-scale renewable energy applications such as solar home systems and provide the rest to the local mini-grid as private renewable electricity (RE) producers. The local network operator, the DSO, would purchase the electricity at a fixed FIT. At the same time, the final consumer could purchase electricity from the DSO at a preferential rate, just like any other consumer connected to the mini-grid. The difference between the preferential tariff offered by the DSO and the FIT paid to the private renewable electricity producer has to be provided at the national level by the electricity authority or regulator (Figure 5.4).

Under the 'monopolist approach', so-called rural energy service companies (RESCOs) have a partial monopoly, i.e. they have the exclusive right for energy

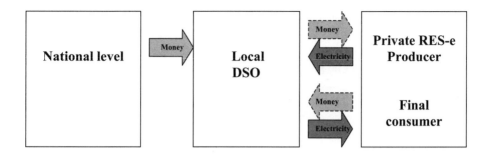

Figure 5.4 *Feed-in tariff for mini-grids (producer/consumer)*

Source: Jacobs, 2009

services and other public services in a given area. They are quasi-governmental organizations, similar to state-owned utilities in monopolistic power markets. They usually provide the full chain of services, including operation, maintenance and repair. Governments frequently offer service concessions to RESCOs for a given period of time of up to 15 years. This permission is based on competitive bidding procedures. In this period, the rural energy service company has the right to exclusively provide all energy services. At the same time, it is obliged to offer the services to everyone who requests them.

The RESCO normally sells electricity to the final consumer below generation costs. This would also be the case if a RESCO incorporates renewable energy plants into its portfolio. In order to guarantee profitability in power generation, the RESCO receives additional money from a higher political level (national or sometimes local) under the FIT scheme (see Figure 5.5). In combination, the FIT payment from the national level and the regulated tariff from the final consumer should offer a sufficient profitability margin. Depending on the specific regulatory design of power markets, the responsible institution is the local energy development agency, the Regulatory Agency or a similar governmental body. This governmental institution is also responsible for the legal and policy framework for renewable energy production, including the setting of the FIT.

Figure 5.5 *Feed-in tariff for mini-grids (RESCO)*

Source: Jacobs, 2009

If the organization at national level does not have sufficient financial means at its disposal, additional sources of income might have to be provided from international donors. In this case, a fund solution for the financing of renewable energy projects should be taken into consideration (see Section 5.3).

5.7 Conclusion

As shown above, FIT schemes have the potential of being designed according to the special needs and conditions of emerging economies and developing countries, including those that may integrate mini-grids, CDM credits and capacity caps into their energy policies. The implementation of FIT schemes in developing countries is especially important as a tool for economic development as many of these emerging economies embark on large-scale projects to bring electricity to their populations. We cannot risk developing countries and emerging economies making the same fundamental mistakes as most industrialized countries have done by basing their energy systems almost exclusively on non-renewable sources. In this light, funding renewable energy projects via FITs in the developing world becomes a task as equally important as promoting them in industrialized countries.

References

ARE (2009) *Best Technology Solution for Rural Electrification*, Presentation at the Sustainable Energy Week, 12 February, Brussels

Bode S. and Michaelowa A. (2003) 'Avoiding perverse effects of baseline and investment additionality determination in the case of renewable energy projects', *Energy Policy*, vol 31, no 6, pp505–517

Jacobs, D. (2009) *Renewable Energy Toolkit: Promotion Strategies in Africa*, World Future Council, May

Moner-Girona, M. (ed.) (2008) *A New Scheme for the Promotion of Renewable Energies in Developing Countries: The Renewable Energy Regulated Purchase Tariff*, European Commission Joint Research Centre, Luxembourg

Reiche, D. and Jacobs, D. (2009) 'Erneuerbare Energiepolitik in Norwegen – Eine kritische Bestandsaufnahme', *Energiewirtschaftliche Tagesfragen*

Schröder, M. (2009) 'Utilizing the clean development mechanism for the deployment of renewable energies in China', *Applied Energy*, vol 86, pp237–242

UNFCCC (2005) *Clarifications on the Consideration of National and/or Sectoral Policies and Circumstances in Baseline Scenarios, (version 02)*, CDM Executive Board, 23–25 November, http://cdm.unfccc.int/EB/022/eb22_repan3.pdf

WHO (2005) *Fact Sheet No 292: Indoor Air Pollution and Health*, World Health Organization, Geneva, www.who.int/mediacentre/factsheets/fs292/en/index.html

6

FIT Developments in Selected Countries

In a way, Feed-in tariffs (FITs) were 'invented' in the US. The Public Utility Regulatory Policies Act (PURPA) was adopted in 1978 as a response to the shortages in oil supply during this decade. PURPA established long-term fixed contracts between qualified independent producers and utilities, obliging the latter to purchase all electricity from the former. In contrast to today's FITs, which usually offer a fixed remuneration which is related to the technology-specific generation cost, the payment under PURPA was based upon the avoided cost of conventionally produced electricity and was non-technology specific. Evaluating best practice after 30 years of experience with FIT schemes led us to include both aspects in our 'bad FIT design' section (see Chapter 4). With the drop in energy prices in the 1990s the avoided costs decreased and thus the remuneration for renewable energy projects followed suit. While in the 1980s about 12,000MW of new renewable energy capacity were installed, the increase in the 1990s was only marginal (Martinot et al, 2006). Even though PURPA can be seen as the first FIT scheme, many design options which are nowadays known to be 'best practice' had not yet been implemented.

At the end of the 1980s, the concept of FITs came to Europe and was further developed. The first FIT schemes were implemented in Portugal (1988), Germany (1990), Denmark (1992) and Spain (1994). Because of the successful deployment of renewable energies at a comparatively low cost in these countries, many other governments decided to implement the same policy mechanism. The international diffusion of this particular support instrument did not come to a halt at the European border. Since the turn of the century, countries and regions all over the world have adopted FIT schemes.

Table 6.1 below gives an overview of where they can be found at the time of writing, at national, state or province level. According to the *Global Status Report 2009*, 45 countries and 18 states/provinces/territories have implemented FITs at some point, amounting to 63 in total (REN21, 2009). The list of countries exploring FITs at the moment is significant, and includes national-level policy in

Table 6.1 *Feed-in tariffs worldwide as of September 2009*[1]

Africa	Americas	Asia	Australasia	Europe
Algeria	Argentina	China	Australia*	Austria
Kenya	Canada*	India*		Bulgaria
Mauritius	Ecuador	Israel		Croatia
South Africa	Nicaragua	Mongolia		Cyprus
Uganda	US*	Pakistan		Czech Republic
		Philippines		Denmark
		South Korea		Estonia
		Sri Lanka		Finland
		Taiwan		France
		Thailand		Germany
		Turkey		Greece
		Ukraine		Hungary
				Ireland
				Italy
				Latvia
				Lithuania
				Luxembourg
				Macedonia
				Malta
				Netherlands
				Portugal
				Slovak Republic
				Slovenia
				Spain
				Switzerland

* Countries with states/provinces which have FITs
Source: Miguel Mendonça and David Jacobs

the US and Australia, as well as many nations, states and provinces in Africa, Asia and the Americas.

In the following sections we will present the latest development of FITs in selected key countries. Due to the constraints of this book we are not able to inform you about the detailed implementation in all countries, states and provinces. But this is not necessary, as we have already given an overview of all basic and advanced FIT design options in the previous chapters.

The purpose of this chapter is to give readers a short overview of the development of FIT schemes in Europe, North America, Australasia and Africa. The European examples of Spain and Germany offer the latest innovations in design options of FIT schemes that have already been in place for more than a decade, due to the larger share of renewable electricity grid and market integration becoming increasingly important. The debate in the UK and North America shows how international policy diffusion can even overcome opposition to renewable energy, at least at the local, state and provincial levels. After years of opposition, the UK government agreed in principle to introduce a FIT scheme for small-scale power

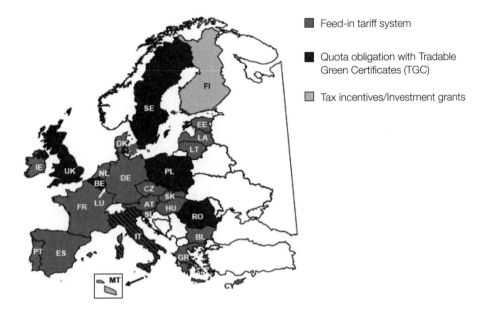

Figure 6.1 *Diffusion of support mechanisms in the EU*

Source: Held, 2008

plants. At the same time, FITs have begun to spread throughout many parts of North America. Finally, the implementation of FITs in Australia, Africa and Asia reveals that renewable energies have ceased to be a niche market.

6.1 Europe

As stated above, the spread of FITs can be especially observed in Europe. 20 of the 27 countries in the EU now operate FITs in order to promote renewable energies. Only six countries operate with tradable green certificate schemes, of which one (Italy) has combined it with a FIT scheme for solar PV. The UK will implement a FIT scheme for small-scale units in the coming years and Finland will introduce a FIT scheme for wind energy (see Figure 6.1).

This diffusion of support mechanisms is partly due to a common European framework as set out by the renewable energy directives in 2001 and 2009. Even though member states in the EU did not agree on a harmonized European support mechanism, the fact that each country had to comply with national targets led to a rapid diffusion of best practice, i.e. well-designed FIT schemes. Germany and Spain in particular are usually considered to be reference cases for other countries, and we will present both countries in the following sections. In addition, we will provide an insight into the FIT debate in the UK, which will implement a FIT scheme for small-scale applications in 2010.

6.1.1 Germany

The first German FIT scheme was enacted in 1990. In the beginning, however, it had many shortcomings, about which you can learn more in Chapter 4. In common with almost all FITs in the 1980s and 1990s, the tariff calculation was based on the avoided costs, tariff payment was linked to the electricity price (see Section 4.7), and the deployment of renewable energy technologies in certain regions was limited by a capacity cap (see Section 4.8). Besides, technology differentiation was insufficient (e.g. solar PV received the same tariff as wind power) and consequently more costly technologies were not promoted. Despite all these shortcomings, the law created substantial growth in the wind power and hydropower sectors. In 1999, the total installed wind capacity reached 4400MW. In the coming years, the input of these branches of the industry was crucial for enacting better legislation to support all renewable energy technologies.

The breakthrough for renewable energies was based on the major revision of the law in 2000, the Renewable Energy Sources Act (EEG). The capacity cap was partly removed and tariffs were calculated based on the generation costs of technologies. Consequently tariff differentiation was better and all technologies received a sufficiently high tariff. Grid connection for renewable energy projects was improved and administrative barriers were removed (see Sections 2.9–2.11). Moreover, the financing mechanism was modified so that all final consumers paid the same throughout the entire territory and interference with European state aid rules could be avoided.

Much of this learning process has influenced the design of FITs all over the world, as the German scheme became the reference model for many other countries or regions. When people speak about the successful German FIT scheme they do in fact refer to this very simple but effective law of 2000 that managed to set out an innovative and stable support framework for renewable energies in only 13 articles. Therefore, any newcomer who seeks inspiration from the German experience should probably take a look at the EEG 2000, as it includes most good design options in a condensed way.

In the following years, the German FIT scheme has been amended twice, in 2004 and 2009.[2] In 2004, the legislator wanted to make further differentiations to tariff payments based on the technology, location and size of the schemes. As the share of renewable energies grew considerably, the policy maker had to take additional measures to avoid windfall profits and control costs for the final consumer. In 2009, this process was continued. Additionally, design criteria for market integration were implemented which allowed for the inclusion of an even larger share of green power in the general electricity mix.

Thanks to the FIT scheme, the share of renewable electricity in the overall electricity mix increased significantly. In 1990 the share of renewable energies in gross electricity consumption was 3.4 per cent. In 2007, it was already more than 14 per cent. In the graph below (Figure 6.2) one can clearly see that, especially

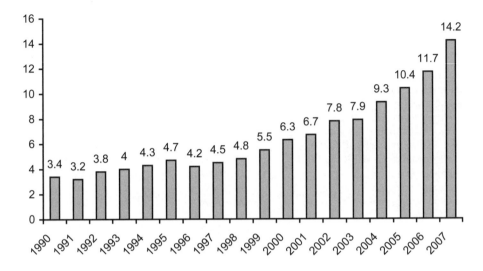

Figure 6.2 *Share of renewable energies in German gross electricity consumption (per cent)*

Source: Adapted from BMU, 2008

from 2000 onwards, the growth rates became more constant due to the favourable support conditions.

This enormous success in promoting renewable energies provoked much momentum in society, the industry and politics. The European target of 12.5 per cent by 2010 was already achieved in 2007. Nowadays, all major political parties in Germany support the FIT scheme. The official target for 2020 is 'at least 30 per cent'. The lead study on renewable energies commissioned by the German Ministry for the Environment includes even more ambitious long-term objectives: more than 50 per cent in 2030, more than 70 per cent in 2040 and almost 90 per cent of final electricity consumption in 2050 (Nitsch, 2008).

Table 6.2 depicts the eligible technologies, the plant size and the tariff level under the current Renewable Energy Sources Act of 2009. The annual degression rates are shown in Section 3.9. The tariff duration period is of 20 years, except for renewed hydropower (15 years).

The German FIT scheme – a cost–benefit analysis

The German case study can be considered to provide impressive figures. In all discussions about FIT schemes, however, sooner or later people want to know what all the costs are. Or, in the German case, what kind of impact does it have on your national economy when you support more than 10 per cent of your electricity via a FIT scheme? And what are the overall benefits?

Table 6.2 *Remunerations under the German feed-in tariff scheme*

Technology	Plant size	Tariff (€/kWh)
Hydropower (new)	Up to 500kW	0.1267
Hydropower (new)	500kW–2MW	0.0865
Hydropower (new)	2–5MW	0.0765
Hydropower (modernized/revitalized)	Up to 500kW	0.1167
Hydropower (modernized/revitalized)	500kW–2MW	0.0865
Hydropower (modernized/revitalized)	2–5MW	0.0865
Hydropower (renewed)	Up to 10MW	0.0632
Hydropower (renewed)	10–20MW	0.058
Hydropower (renewed)	20–50MW	0.0434
Hydropower (renewed)	Over 50MW	0.035
Landfill gas	Up to 500kW	0.09
Landfill gas	500kW–5MW	0.0616
Sewage gas	Up to 500kW	0.0711
Sewage gas	500kW–5MW	0.0616
Mine gas	Up to 1MW	0.0716
Mine gas	1–5MW	0.0516
Mine gas	Over 5MW	0.0416
Biomass	Up to 150kW	0.1167
Biomass	150–500kW	0.0918
Biomass	500kW–5MW	0.0825
Biomass	5–20MW	0.0779
Geothermal	Up to 10MW	0.16
Geothermal	Over 10MW	0.105
Onshore wind	Location-specific tariffs	0.0502–0.092
Offshore wind	Location-specific tariffs	0.035–0.13
Solar radiation (roof-mounted)	Up to 30kW	0.4301
Solar radiation (roof-mounted)	30–100kW	0.4091
Solar radiation (roof-mounted)	100–1000kW	0.3958
Solar radiation (roof-mounted)	Over 1000kW	0.33
Solar radiation (free-standing)	All	0.3194

Source: Adapted from BGB, 2008

The cost–benefit analysis is an important part of the progress report of the German EEG. The last progress report was issued in 2007 (BMU, 2007). When analysing the following figures, one has to keep the success of the German FIT scheme in mind: the share of renewable electricity was increased from 6.3 per cent in 2000 to more than 14 per cent in 2007 (71.5TWh)!

The overall tariff payment increased from €3.6 billion in 2004, to €5.8 billion in 2006 and €8.9 billion in 2008 (BMU, 2007). Even though these figures seem huge at first glance, what matters to the final consumer is the cost difference with conventional power production. The German Ministry for the Environment evaluates the so-called differential costs, i.e. the difference between the costs for conventional power generation and renewable power generation under the EEG. These costs increased as well, but at a much slower pace due to the cost increase for fossil fuels. The differential costs amounted to €2.5 billion in 2004, €3.3 billion in 2006 and €4.5 billion in 2008 (BMU, 2007; IfnE, 2009). According to estimates, these costs will further increase to €5.4 billion in 2015 before they go down to €4.6 billion in 2020 and €0.6 billion in 2030 (Wenzel and Nitsch, 2008).

From 2015 onwards the additional costs for renewable electricity generation will decrease. This is partly due to lower costs for renewable electricity generation and also to increases for conventional power production. Consequently, renewable energy deployment today will have a stabilizing effect on tomorrow's electricity price. In 2008, the average power price on the German power exchange increased significantly. Partly because of the European Emissions Trading Scheme, the spot market price went from €0.038/kWh in 2007 to €0.066/kWh in 2008. In 2009, an average price of €0.068/kWh can be expected. The average cost of one kilowatt hour of electricity produced under the FIT scheme is €0.12: €0.076/kWh for hydro, €0.07/kWh for biogases, €0.14/kWh for biomass, €0.15/kWh for geothermal, €0.0877/kWh for wind power and €0.51/kWh for solar PV (IfnE, 2009). For each final consumer, the additional costs due to renewable electricity support remains relatively low. According to the last calculation, each household (3500kWh per year) pays around an additional €3 per month. This amount will probably increase to a maximum of €4–4.5 per month in 2015.

In 2006, the total costs of the German FIT scheme amounted to €3.3 billion, together with minor costs for regulation energy (€0.1 billion) and additional transaction costs for grid operators (€0.002 billion). In the same year, the FIT scheme generated significant benefits, related to the reduction in the wholesale electricity price, the avoided external costs for electricity generation and the avoided energy imports. As we will see, the benefits from the FIT scheme were much larger than the costs.

The reduction of the wholesale electricity price is largely due to the so-called merit-order effect (Bode, 2006; da Miera et al, 2008; Sensfuß et al, 2008). When decisions are made about electricity dispatch, i.e. which plants are to provide power to cover electricity demand, the power production units are ordered according to their variable costs, the so-called merit order. Large shares of cost-effective renewable power, as provided by wind energy, can significantly reduce the spot market electricity price. This is due to the fact that supply companies will have to buy the renewable electricity in advance and therefore the remaining demand is reduced. As shown in Figure 6.3, this reduction in demand led to a decrease of the electricity price as shown by the dark grey area. The market price of electricity is determined by the last, most expensive unit of electricity to match supply and demand.

In Germany, this effect has reduced the average market price of electricity by €7.83/MWh, leading to overall savings of almost €5 billion in 2006 alone. In Spain, the same effect reduced the market price by €6/MWh and saved utilities and consumers €1.7 billion, again more than the total costs of tariff payment. It has to be stated that the merit-order effect is not directly related to FITs. The effect

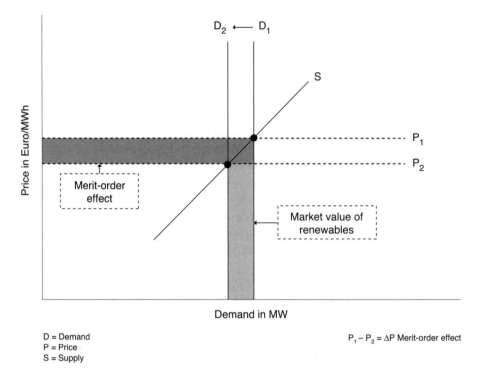

Figure 6.3 *The merit-order effect*

Source: Sensfuß et al, 2008

might also occur under other support mechanisms which enable large amounts of electricity to be produced from renewable energy sources in a cost-effective way.

Besides this reduction in the wholesale price, renewable energies replace conventional fuels and thus avoid negative external costs for electricity from these power sources. In 2006, the German FIT scheme avoided 45 million tonnes of CO_2 emissions which otherwise would have damaged the environment. According to estimates this saved the German economy €3.4 billion in 2006. Moreover, energy dependency was significantly reduced and high costs for the import of hard coal and gas were avoided. This led to additional savings of €1 billion. All the above-mentioned costs and benefits of the German FIT scheme in 2006 are depicted in Table 6.3.

As a result, the benefits of the FIT scheme have long surpassed the costs for the German economy as a whole. Besides, the German FIT scheme delivers several other positive social and environmental effects, thus creating a classic win–win–win situation in economic, ecological and social terms. In 2008, almost 280,000 people were employed in the renewable energy industry, an increase of 12 per cent compared to 2007. Of these employees, at least 150,000 can be attributed to the German FIT scheme. The industry estimates that by 2020 at least half a million

Table 6.3 *Costs and benefits of the German FIT scheme (€ billion, 2006)*

Costs		Benefits	
EEG differential cost	3.2	Reduction in the wholesale price (merit-order effect)	5.0
Additional costs, regulation energy	0.1	Avoided external costs, electricity generation	3.4
Transaction costs	0.002	Avoided energy imports	1.0
Total costs	3.3	Total benefits	9.4

Source: IfnE/BMU, 2007; Sösemann/BMU, 2007

people will work in the renewable energy sector. The German industry already has a total turnover of €30 billion today. In addition, greenhouse gas emissions were reduced by 115 million tonnes in 2008, of which 57 million tonnes can be directly related to the German FIT scheme.

6.1.2 Spain

Spain first implemented a full-scale FIT in 1994. However, the basic principles for renewable energy support were laid out in the law for energy conservation in 1980, including guaranteed grid access, contracts for power purchasing and certain price guarantees (Dinica and Bechberger, 2005). Even though the decree of 1994 established a minimum purchase duration of five years, and offered technology- and plant-size-specific tariffs, the early Spanish FIT scheme was of limited effect. Despite the fact that administrative barriers and grid access were improved in 1998, the major shortcoming was the lack of investment security. Up until 2004, the tariff level could change every year, thus making the financing of renewable electricity projects largely unpredictable (Jacobs, 2008).

The legislative breakthrough came in 2004, with the Royal Decree 436/2004 and the amendment in 2007 (BOE, 2007). By 2003, the Spanish regulator CNE had already established a transparent methodology for calculating tariffs based on their average generation costs (see Section 2.3). From 2004 on, tariff payment was guaranteed for at least 15 years, in the case of some technologies even for the whole lifetime of a power plant. In order to better integrate renewable electricity into the grey power market, all units with an installed capacity of more than 10MW had to make provision for future power production (see Section 3.5). In 2007, the tariff payment was eventually disconnected from the average electricity price and based on a fixed tariff payment per kilowatt-hour. Additionally, the tariff level was increased for a large number of technologies which were still far from reaching the 2010 mid-term targets, including biomass and solar thermal.

One of the best-known design options of the Spanish FIT scheme is the so-called premium FIT. Instead of receiving a fixed tariff payment, the producer can decide to sell the electricity on the spot market and receive a smaller feed-in tariff payment on top (see Section 3.1). Under the 1998 legislation, producers already had the chance to choose between both options. However, only under the 2004 legislation did the premium FIT option become economically attractive, as the expected returns on investment were a few per cent higher then under the fixed tariff payment option. The premium FIT is especially popular among wind power producers. At the start of 2005, only 20 per cent of all wind power producers sold their electricity under the premium feed-in tariff option. By the end of the same year, 98 per cent had already selected this sales option. This figure has remained relatively stable for the last couple of years (Jacobs, 2008). The establishment of a cap and a floor in 2007 helped to avoid windfall profits for producers under the premium FIT (see Section 3.1).

Table 6.4 indicates the eligible technologies, the plant size and the tariff level for both the fixed-price option and the premium FIT option. The duration of the tariff payment is 25 years for solar and hydropower, 20 years for geothermal, wave, tidal, hot dry rock and ocean current energy, and 15 years for all types of biomass

Table 6.4 *Tariff payment under the Spanish feed-in tariff scheme*

Technologies	Plant size	Fixed tariff (€/kWh)	Premium tariff (€/kWh) – including cap and floor
Onshore wind	Up to 50MW	0.073	0.029 (0.071–0.085)
Offshore wind	Up to 50MW	–	0.084 (0.084–0.164)
Geothermal; wave; tidal; hot dry rock; ocean current	Up to 50MW	0.069	0.038
Biomass (energy crops)	Up to 2MW	0.158	0.115 (0.154–0.166)
Biomass (energy crops)	2–50MW	0.147	0.101 (0.143–0.151)
Biomass (agricultural waste)	Up to 2MW	0.126	0.082 (0.121–0.133)
Biomass (agricultural waste)	2–50MW	0.107	0.062 (0.104–0.112)
Biomass (forestry material)	Up to 2MW	0.126	0.082 (0.121–0.133)
Biomass (forestry material)	2–50MW	0.118	0.073 (0.114–0.123)
Biogas (landfill)	Up to 50MW	0.080	0.038 (0.074–0.09)
Biogas (others)	Up to 500kW	0.131	0.098 (0.124–0.153)
Biogas (others)	500kW–50MW	0.097	0.058 (0.096–0.11)
Biogas (liquid biomass)	Up to 50MW	0.053	0.031 (0.051–0.083)
Hydropower	Up to 10MW	0.078	0.025 (0.065–0.085)
Hydropower	10–50MW	–	0.021 (0.061–0.08)
Solar PV (roof-mounted)	Up to 20kW	0.34	–
Solar PV (roof-mounted)	Over 20kW	0.32	–
Solar PV (free-standing)	Up to 50MW	0.32	–
Solar thermal	Up to 50MW	0.269	0.254 (0.344–0.254)

Source: Adapted from BOE, 2007, 2008

and biogas. After the indicated number of years, renewable energy plants continue to receive a reduced tariff payment. The level, however, is often in the range of market prices for electricity.

In Spain, the installed renewables capacity increased significantly compared with previous years. While the installed wind power in 1997 was of only 440MW it reached more than 15,000MW in 2007. However, due to the lack of policy measures for improved energy efficiency and energy savings, Spain has difficulties reaching its international targets and obligations. In line with the European renewable electricity directive of 2001, Spain will have to increase the share of green power to 29.4 per cent. The new directive of 2009 obliges Spain to increase the share of renewable energies in total energy consumption to 20 per cent. According to estimates from the Spanish wind energy association AEE, this translates into a 45 per cent share in the Spanish electricity market. In no more than 11 years, Spain will already produce almost half of its power from renewable energy sources!

Despite the successful deployment of renewable energies, there is still room for improving the Spanish FIT scheme. First and foremost, the installed capacity is capped. On the one hand, the Spanish scheme only covers installations of up to 50MW. This limit is due to the overall framework for renewable energies under the so-called 'special regime'. In the past, it was believed that renewable energy power plants are by definition small-scale. Nowadays, it is clear that wind power plants, solar thermal plants and other technologies can easily exceed this limit. On the other hand, the technology-specific mid-term target as expressed by the national renewable energy plants can hinder the growth of renewable energies as tariff payment for new installations can come to an end. This is especially critical for the solar PV industry. Besides, the Spanish FIT scheme has the legal rank of a Royal Decree. Even though it is 'stronger' than for instance a ministerial order, the Spanish renewable energy associations have long called for a FIT law. Before the last general elections, the current Socialist government had promised to initiate the respective legislative process, but up to now nothing has changed.

6.1.3 The UK

Many years of UK government opposition have been removed by a very determined 'tariff coalition', aided by governmental changes in leadership, responsible institutions and, apparently, priorities. After a fascinating campaign, the government added a provision to the Energy Act of 2008 for a FIT for small-scale renewables generation (up to 5MW). They have committed to introduce a FIT for electricity in April 2010 and for renewable heat in April 2011.

The coalition produced a Tariff Blueprint Document.[3] This comes from the Renewable Energy Association (REA), the UK's main renewables trade association, which coordinated working groups around various elements of the FIT for electricity. This first helped to develop an understanding of the needs and wants of the various stakeholders; second, it established some approaches to designing and setting tariffs; and, third, it provided a benchmark to keep a notoriously uninterested government on its toes.

It is also interesting to look at the UK because work is being done with fuel poverty groups on ensuring that the rise in electricity costs from renewables deployment does not affect the poorest members of society through their energy prices. This is particularly pertinent in a nation with some of the highest retail energy prices in Europe.

Summer 2009 saw the latest in a long line of government consultations on energy. Rather than go over the largely unpleasant details of their nuclear consultations, we can summarize by saying that they have apparently developed a reputation in recent years for being one-sided paper exercises with foregone conclusions (for the unpleasant details see Girardet and Mendonça, 2009; or contact Greenpeace UK).

By taking the initiative, the tariff coalition groups stand a very good chance of ensuring that the Department of Energy and Climate Change (DECC) cannot easily follow in the obstructive footsteps of one of its 'parent' departments, the Department of Trade and Industry (DTI). If our attitude towards the government's performance is somewhat jaundiced, it arises from innumerable media reports, conversations with a multitude of individuals from business, non-governmental organizations (NGOs), academia, journalism and politics – and personal experience.

Despite this, the creation of DECC, and the appointment of Ed Miliband as Secretary of State for Energy and Climate Change, saw a rapid shift from an impassable 'no' to a qualified 'yes' on a FIT. Their main concern seems to have been that the current system for large renewables, the Renewables Obligation (RO), should remain unaffected by the establishment of another national support scheme. This should be the case as the proposed FIT has a 5MW cap, and the RO is said to be of limited value for projects below 20MW.

The RO has been fiercely defended by the large energy companies – the 'big six' who monopolize the UK's energy market (some of whom are owned by German companies who have fought the German FIT for many years). The RO has been fiercely criticized for costing too much and producing too little. It also keeps the market in the hands of the existing energy producers; when added to the prices it pays, it is clear why it would be defended with such determination (discussed further in Chapter 10).

The design and implementation of the FIT is being handled by ORED, the Office for Renewable Energy Deployment (ORED), housed within DECC. ORED is responsible for pursuing the Government's Renewable Energy Strategy (published in July 2009). Their responsibilities include support scheme development, overcoming the non-financial barriers (such as planning issues), promoting sustainable bio-energy and wave and tidal power, addressing supply chain blockages, and developing business opportunities for the UK renewable sector.

Current government proposals, as of September 2009, are encouraging, and seem relatively simple and clear. Wisely, they did not attempt to re-invent the wheel but instead base the proposal for their FIT on international best practice. The proposed system features a fixed payment per kWh generated, but with an extra payment per kWh exported to the grid. There is a further incentive for on-site use. Every eligible technology will get 20 years, other than PV, which will get 25 years of payment.

Tariff degression will feature for new installations. Degression rates will vary for each technology, going up to 7 per cent in the case of solar PV. There is some discussion, naturally, about balancing initial rates and degression rates, and the UK example is interesting because the market is still very small compared to the potential. Where a similar degression rate may work in a more mature market, in the UK a better initial rate for technologies would be useful to get things moving. Once the market is beginning to expand and become evident in the urban and rural landscape, degression rates are more useful for controlling costs. As manufacturing has yet to be established, let alone producing significant economies of scale, the UK deployment of microgeneration will not lead to dramatic cost reductions in the near future. Initially, the technologies will be imported from developed markets.

The feed-in tariff scheme includes a wide range of technologies – namely, wind, solar PV, hydro power, anaerobic digestion, biomass, biomass CHP and non-renewable micro CHP. Interestingly, the UK FIT is one of the first support mechanisms offering a special tariff level for micro wind energy systems. In the case of PV, the FIT scheme clearly focuses on roof-mounted PV installation, offering a significantly lower tariff for stand-alone systems. Government officials have suggested that more technologies might become eligible at a later stage. The tariffs are calculated based on the technology-specific generation costs, offering producers a return on investment between 5 and 8 per cent.

However, there are various concerns expressed by stakeholders, especially small-to-medium enterprises in the industry. They wish to see improved rates and return on investment for some technologies, and have the eligibility backdated for existing installations. Civil servants argue that one would not expect a rebate from Apple just because one buys an iPod the day before they drop the price. 'No free lunch' was the phrase employed. On the rates, they make the same point we do: that a balance must be struck between investor confidence and public support.

The REA report suggests that the tariffs will not necessarily be perfect from day one, but 'roughly right', and would prefer to set them on the high side to begin with, in order to learn more. If they are set low and nothing happens, this provides less indication of what will work most efficiently than would a high tariff with high take-up. A recent study suggested that with the right incentives in place, the UK could see installations of several hundred MW of photovoltaics annually within three years (Martin, 2009). To address fundamental issues such as tariff levels, take-up and costs, the scheme will include a first review of the tariffs after one year, allowing amendments if necessary. Further amendments will take place every three years thereafter.

In the past, The UK government has not established confidence in the domestic renewable energy industry, and has even been implicated in attempts to wreck EU target-setting or reduce its commitment to actually build significant amounts of new capacity (Seager and Gow, 2007). However, the EU's targets are now established and the UK is compelled to meet them. The FIT may become instrumental in making the necessary progress.

6.2 North America

Parts of Canada and the US have begun to implement FITs, but not at the national level. Instead, the province of Ontario has the continent's most well-developed FIT, and the city of Gainesville in Florida and the state of Vermont both passed the first true FITs in the US earlier this year. Policies are developing in other parts of the continent as momentum seems to build in favour of FITs.

6.2.1 Canada

There is no national FIT in Canada and there is unlikely to be one in the traditional sense because of Canada's unique political history. Energy and especially electricity policy is the domain of each province. There have been attempts in several provinces with little success. However, in mid-2009 the province of Ontario implemented the first true system of differentiated FITs in North America. The policy has been five years in the making and not without its setbacks, but recent changes appear to have improved things considerably.

The original Renewable Energy Standard Offer Program (RESOP) was a simple FIT programme launched in November 2006. It provided a fixed rate of CA$0.11/kWh for hydroelectric, biomass and wind projects and CA$0.42/kWh for solar PV facilities up to 10MW in size. It set these rates in 20-year contracts and provided access to the grid. In the first 15 months of the programme, all went well: more than 1300MW of renewable energy contracts were signed and Ontario was on its way to becoming a regional leader in renewable energy deployment.

Then things changed. The Ontario Power Authority (OPA), responsible for managing the provincial FIT, started limiting the amount of renewable energy that could be connected to the grid in certain regions of the province. The OPA announced plans to construct 14,000MW of new nuclear capacity over the next two decades to wean the province off of fossil fuels and replace the ageing reactors already in place. The OPA decided to revise key aspects of the FIT in 2008. The OPA placed a moratorium on all new renewable power contracts over 10kW for many parts of Ontario and limited developers to one 10MW project per transformer station across the entire service area (Sovacool, 2008).

Thankfully, things changed in mid-2008 with the appointment of a new Minister of Energy, George Smitherman, who was given a mandate to restart the renewable energy programme to meet plans for closing the remainder of the province's coal plants by 2014. The Minister of Energy was also the Deputy Premier, the second highest office holder in the province, and the appointment indicated the seriousness with which the Premier considered promoting renewable energy.

Following the Minister of Energy's participation in the World Wind Energy Association's 2008 conference in Kingston, Ontario, events began to move fast. The Minister of Energy subsequently toured renewable energy projects in Denmark, Germany and Spain. When he returned he decided to create an entirely new programme to emulate European success and make Ontario the leader in renewable energy development in North America.

The resulting Green Energy and Green Economy Act (GEA) became law on 14 May 2009. The multifaceted law required new programmes for energy conservation and authorized the Minister of Energy to call for new sources of renewable energy through a system of FITs. His directive to the OPA was intended to remove many of the former restrictions and implement a full system of differentiated tariffs (see Table 6.5). Apart from bolstering the OPA's FIT scheme, the GEA also set voluntary requirements for home energy audits prior to the sale of homes, streamlined permitting and siting approvals for renewable energy projects, and established an academic research centre to investigate the public health impacts of renewable energy projects.

According to the OPA, the revised tariff schedule elicited responses from more than 150 developers representing 381 projects amounting to 15,128MW of potential supply (Ontario Power Authority, 2009). Many of these projects are in the small to medium scale and represent a diverse array of technologies, with 266 projects constituting 10kW–10MW of capacity (1657MW in total) and 115 projects between 10.1MW and 600MW (13,470MW in total) (see Table 6.6). Because of the long-term certainty offered by the Ontario FITs, Everbrite Solar announced in 2009 that they intended to build a 150MW thin-film manufacturing facility in Kingston, Ontario, a sign that the market for renewable energy is indeed growing quickly there (Farrell, 2009). The FIT scheme is popular with the public as well, with 87 per cent of respondents in a recent poll indicating that they supported FITs and widespread deployment of renewable energy (Green Energy Act Alliance, 2009). These results were backed by yet another recent report estimating that the GEA is likely to create 90,000 jobs per year in the energy efficiency, renewable energy, and transmission and distribution sectors (Green Energy Act Alliance, 2009).

The FITs promoted by the GEA and OPA have thus put Ontario at the forefront of renewable energy policy in North America. Ontario's approach may well become the model emulated across the continent. Some are calling the Ontario programme the most progressive renewable energy policy in North America in more than two decades. Not since the US Congress passed PURPA in 1978 has a single policy had the potential for such wide-ranging influence on energy policy as Ontario's Green Energy Act.

Table 6.5 *The Ontario Power Authority's revised FITs, May 2009*

	Years	1.586 €/kWh	CA$/kWh	0.860 US$/kWh
Wind				
Onshore*	20	0.0851	0.135	0.116
Offshore	20	0.1198	0.190	0.163
Photovoltaics				
Rooftop or ground-mounted < 10kW	20	0.5055	0.802	0.690
Rooftop >10kW <250kW	20	0.4494	0.713	0.613
Rooftop >250kW <500kW	20	0.4003	0.635	0.546
Rooftop >500 kW	20	0.3398	0.539	0.464
Ground-mounted <10 MW*	20	0.2792	0.442	0.381
Hydro				
<10MW*	40	0.0826	0.131	0.113
>10MW <50 MW*	40	0.0769	0.122	0.105
Landfill Gas				
<10MW*	20	0.0700	0.111	0.095
>10MW*	20	0.0649	0.103	0.089
Biogas				
<500kW*	20	0.1009	0.160	0.138
>500kW <10 MW*	20	0.0927	0.147	0.126
>10MW*	20	0.0656	0.104	0.089
Biomass				
<10MW*	20	0.0870	0.138	0.119
>10MW*	20	0.0819	0.130	0.112

Notes: *Eligible for Aboriginal or Community Bonus.
Inflation adjustment: 100% during construction (3 years), 20 per cent during life of contract.
Source: www.powerauthority.on.ca/fit/Storage.asp?StorageID=10143

Table 6.6 *Expression of interest in Ontario Power Authority's FITs*

Fuel type	Number of Projects	Aggregate Capacity (MW)
Wind	164	13,382
Hydro	58	374
Solar PV	121	1213
Bioenergy	38	159
Total	381	15,128

6.2.2 US

The situation in the US is more complex than in Canada. Fixed-price incentives for renewable energy have been around since the 1970s, although these early programmes do not closely resemble modern FITs. The two most commonly mentioned historical policies are PURPA, enacted in 1978, and standard offer contracts in California.

PURPA was one of five statutes that were included in President Jimmy Carter's National Energy Plan as an attempt to reduce US dependence on foreign oil and vulnerability to supply interruptions, and to develop renewable and alternative sources of energy.[4] After the passage of PURPA, electricity suppliers were no longer able to hold a monopoly over power generation. PURPA enabled new actors, such as small power producers or 'qualifying facilities', to generate electricity on their own and forced the incumbent utilities to purchase this power at a reasonable fixed rate based on the 'avoided costs' to the utility. PURPA was a breakthrough in the sense that it opened the door to non-utility producers of power, although it did not catalyse widespread use of renewables because the 'avoided costs' were still too low, often ranging from a mere $0.02–0.05/kWh. Despite its limitations, PURPA was perhaps the first major piece of legislation to offer a fixed payment to small-scale renewable power producers, and from 1980 to 1992 (before the next major legislative act relating to electricity was passed) about 40,000MW of non-utility generating capacity was added to the country's grid (Edison Electric Institute, 1996). The state of California, for example, implemented PURPA through standard offer contracts that saw the addition of 1200MW of wind capacity between 1984 and 1994 (Gipe, 1995).

Other states and utilities throughout the country have since experimented with 'performance-based incentive payments' to promote renewable electricity. Minnesota passed its 'Community-Based Energy Development Proposal' in 2005 to allow utilities to give wind projects within the state $0.055/kWh, and the state of Washington signed a solar PV programme into law that pays as much as $0.54/kWh to produce solar electricity. California piloted a modest PV tariff in 2005 of $0.5/kWh (funded out of their system benefits charge), and Wisconsin, Vermont, and the Tennessee Valley Authority offer various types of fixed tariffs as part of their green power programmes at the utility scale (Gipe, 2006; Rickerson and Grace, 2007; Grace et al, 2008, 2009; Cory et al, 2009; Couture, 2009).

These state programmes, however, do not have the key components that other successful FIT schemes do. Many are not based on the costs of renewable energy generation and do not offer rates high enough to make investments in renewable energy profitable. Most set caps on project size or cost. The majority do not differentiate tariffs by size of the project or type of technology. They are usually voluntary and do not guarantee access to the grid. And, crucially, they do not spread costs of the tariff among all customers, instead spreading it only among those willing to pay a premium. The Minnesota tariff for wind energy, for

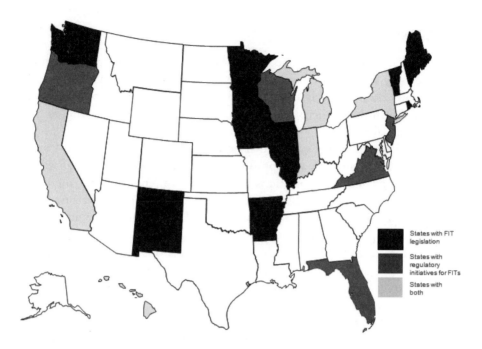

Figure 6.4 *Locations in the US with FIT legislation and/or regulatory initiatives (as of May 2009)*

Source: Wilson Rickerson

example, was initially limited to 100MW, capped project size at 2MW, did not have components guaranteeing interconnection and priority grid access, and did not mandate that utilities have to offer it.

That said, there is progress among the individual states to adopt FITs that more closely resemble those in Europe and elsewhere. As of May 2009, discussions for comprehensive FIT programmes at the legislative or regulatory level were occurring in no fewer than 18 states: Arkansas, California, Florida, Hawaii, Illinois, Indiana, Iowa, Maine, Michigan, Minnesota, New Jersey, New Mexico, New York, Oregon, Rhode Island, Vermont, Virginia and Washington (see Figure 6.4).

By May 2009, the only formal FITs in the US meeting our criteria were in Gainesville, Florida and in the state of Vermont. In Florida, the board of directors for the regional utility, the Gainesville City Commission, unanimously approved the creation of 'Solar Energy Purchase Agreements' in February 2009. The Gainesville FITs give eligible small solar projects (below 25kW) $0.32/kWh for the electricity they export to the grid and larger ground-mounted projects (greater than 25kW) $0.26/kWh, and they guarantee the rate for 20 years. Under the programme, Gainesville Regional Utilities will purchase all of the electricity produced by these systems and then sell it back to residential and commercial customers for $0.12/kWh. The tariff of $0.32/kWh was designed to give investors in solar energy a 5 per cent return on investment for larger projects. The difference in cost between the two tariffs will be paid for by all Gainesville utility customers, and it is expected that the extra costs will not exceed $4–$5 per month, less than a large cup of Starbucks coffee (Gainesville Regional Utilities official, pers comm May 2009). The only caveat is that the Gainesville FIT does set a cap on total installations at 4MW per year, and even though the utility only serves 90,000 customers, the FIT has already been fully subscribed (Morris, 2009; Rickerson, pers comm May 2009). According to officials at Gainesville Regional Utilities, by April 2009 the utility had received applications for more than 40MW of solar PV capacity in their service area and have more or less booked projects through to 2012 (Couture, pers comm May 2009).

Vermont became the first state to implement a full system of FITs in late May 2009, when they passed legislation (H. 446) altering the state's Sustainably Priced Energy Enterprise Development Program, or SPEED. The changes to SPEED provide FITs intended to cover generation costs plus a reasonable profit, with the costs of the programme distributed among Vermont electricity ratepayers. The Vermont FITs provide long-term contracts for 20 years and provide a specific tariff for small-scale wind turbines less than 15kW (See Table 6.7). Tariffs are differentiated by technology and size, and the programme will be reviewed periodically. The FIT legislation instructs the Vermont Public Service Board to review and reset the tariffs every two years to keep the programme efficient. The Executive Director of Renewable Energy Vermont, one of the groups that campaigned for the FITs, argued that 'this law puts Vermont in a leadership role on renewable energy policy and will help to bring vibrant growth and development to our local renewable energy industry' (Gipe, 2009).

State and city action has so far not been matched by serious commitment in the US Congress for a FIT at the federal level. In March of 2008, Representative Jay Inslee from Washington introduced a proposal for a national FIT under legislation named the 'Clean Energy Buy-Back Act' which was later renamed the 'Renewable Energy Jobs & Security Act' (HR 6401) in June 2008. Inslee's proposal was backed

Table 6.7 *Vermont's recently enacted FITs*

Programme cap	50MW
Project size cap	2.2MW
Contract term	20 years
Rate of return	Profit set at same rate of return for Vermont electric utilities
Programme evaluation	Every two years
Specific tariffs:	
Wind energy <15kW	$0.2/kWh
Wind energy >15kW	$0.14/kWh
Landfill and biogas	$0.12/kWh
Solar	$0.3/kWh

Source: Gipe, 2009

by more than 70 renewable energy companies and organizations, but never even passed committee in the US House of Representatives. Another federal bill was introduced in 2009 but would create tariffs *below* the avoided cost of generating electricity, hardly a FIT by our standards. Hopefully, as more cities and states enact their own FITs in the US, that will start to change.

6.3 AUSTRALASIA

The antipodean nations of Australia and New Zealand are both engaged in efforts to promote renewable energy. Australia is further ahead in terms of policy implementation, and has many state-level schemes in operation. However, a truly effective national-level policy is a matter of debate, as is the use of either a 'net' or 'gross' FIT. New Zealand has a FIT campaign in development, which at the time of writing is facing opposition from a monopoly situation, wherein the government is part-owner of the major energy producer.

6.3.1 Australia

Australia is one of the few FIT countries that does not have a nationally consistent FIT, primarily due to the nature of the grid system in Australia and the jurisdiction responsibilities of state governments. The FIT policies that are currently in place or planned in the different states and territories differ in payable rates, eligible renewable energy sources, requirements of distributors and retailers, and so on.

Australia has committed to achieving 20 per cent of its electricity generation from renewable sources by 2020. Various incentive programmes have therefore been put in place by the Commonwealth government and individual state governments to promote growth and provide support to the industry, which includes state-level FITs. Many states and territories have either implemented, or committed to, a state-based FIT system. The purpose is to encourage the uptake of small-scale renewables, but solar PV systems under 30kW are typically being targeted at the time of writing. Many would like to see more technologies being supported, including wind, bioenergy and even large-scale power systems such as large-scale solar. There are challenges, however, such as recognizing that small-scale bioenergy, which has a significant collective generation potential, is not economically viable without additional support. Australian rural and regional communities and their enterprises in general could be boosted by the many possibilities in these other renewables that can be encouraged through FIT:

- Most of the state-based FITs that have been implemented or announced are 'net'. These schemes only pay for the excess electricity exported (production minus consumption) to the grid, instead of the total 'gross' generation. This

difference can have a significant impact on the incentive received and therefore the payback period of the renewable energy system.
- There has been considerable debate around net and gross FIT schemes. These voices of support and opposition have come in the form of protest rallies and online petitions in various states. There are also system installers requesting dispensation in the FITs to allow switching from one scheme to another when the connection arrangements for either one of them is not cost-effective. Australian Capital Territory is currently the only state that is actively offering a gross FIT, and the Western Australia government has indicated that a similar scheme is imminent.

In June 2008, the Senate referred the Renewable Energy (Electricity) Amendment (Feed-in Tariff) Bill 2008 and brought together the Senate Environment, Communications and the Arts Committee to conduct an enquiry into establishing a national FIT scheme. Responses to the request for submissions to that enquiry were received from 129 organizations.

In November 2008, the Council of Australian Governments (COAG) agreed to investigate the harmonization of FIT legislations. COAG then set out 'National Principles for Feed-in Tariff Schemes'. These principles do not appear to support the implementation of a gross FIT but do seek to streamline the state-based FITs to be nationally consistent. The progress of its implementation is subject to further announcements. The report produced by Access Economics for the Clean Energy Council demonstrated that the difference in supporting the solar industry with a gross FIT, as opposed to a net one, can be significant. In the short term, a substantial shift from a net to a gross FIT in the different states and territories does not seem to be likely but this may change in the long term.

6.3.2 Australian state policies

Australian Capital Territory

The Australian Capital Territory FIT scheme is legislated under the Electricity Feed-in (Renewable Energy Premium) Act 2008 and came into effect in March 2009 (see Table 6.8). The Act mandates a review of the scheme at least once every five years, when the premium rate may be adjusted. The Australian Capital Territory currently offers the most generous FIT in Australia which is a paid premium of 3.88 times the market rate (i.e. AU$0.5005) for 20 years. It is also the only state that is actively offering a gross FIT.

Table 6.8 *Australian Capital Territory feed-in tariff details*

Date of first implementation	1 March 2009
Types of eligible renewables	Solar, wind and any other source determined by the minister
Sectors eligible	Domestic and commercial
Purchase obligation	Up to 10kW capacity: AU$0.5005/kWh
	Up to 30kW capacity: AU$0.4004/kWh
Scheme type	Gross
Generation cap	Not applicable
Capacity cap	30kW
Duration of scheme	20 years
Calculation methodology	Electricity price

New South Wales

In November 2008, it was announced that a taskforce was to be established to investigate and propose a FIT for New South Wales. It is anticipated that the state government will implement a scheme that is in line with the proposed national guidelines by COAG, which is a FIT harmonized with other schemes currently operating in other states and territories (see Table 6.9). In March 2009, the state government announced its commitment to implement the scheme towards mid 2009.

Table 6.9 *New South Wales feed-in tariff details*

Year of first implementation	To be confirmed (TBC)
Types of eligible renewables	TBC
Sectors eligible	TBC
Purchase obligation	Estimated to be AU$0.6/kWh
Scheme type	TBC
Generation cap	TBC
Capacity cap	TBC
Duration of scheme	TBC
Calculation methodology	TBC

Northern Territory

Northern Territory does not currently have a consistent FIT throughout the territory and one of the complexities of the FIT implementation is the nature of their micro-grid networks. Northern Territory manages a 225 Rooftop Photovoltaic Systems programme that is a gross FIT scheme for the Alice Springs Solar City (Table 6.10). As part of the federal government's Solar City project, Alice Springs in Northern Territory was selected as one of the four locations to deploy energy efficient measures with some emphasis on solar power. The price of the FIT rate is currently set at more than double the retail price which is capped at AU$5 per day at a rate of AU$0.4576/kWh for residential generation and AU$0.32/kWh for commercial generation.

Table 6.10 *Northern Territory feed-in tariff details*

Year of first implementation	2007
Types of eligible renewables	Solar
Sectors eligible	To be confirmed (TBC)
Purchase obligation	AU$0.4576/kWh
Scheme type	Net
Generation cap	Individual system: AU$5 per day, then reverts to AU$0.2311/kWh
Capacity cap	TBC
Duration of scheme	TBC
Calculation methodology	TBC

Queensland

The Solar Bonus Scheme by the Queensland government came into effect in July 2008 to provide a net FIT for new and existing residential and small business energy generators (see Table 6.11). The legislation requires the retailers to purchase the surplus electricity exported into the grid at a rate of AU$0.44/kWh which is approximately three times the current retail price. This FIT scheme will be offered until 2028 and will either be reviewed after ten years of implementation or when 8MW of installed capacity is reached, whichever comes first.

Table 6.11 *Queensland feed-in tariff details*

Date of first implementation	1 July 2008
Types of eligible renewables	Solar
Sectors eligible	Domestic and small business
Purchase obligation	AU$0.44/kWh
Scheme type	Net
Generation cap	100MWh per year
Capacity cap	10kW – single phase connection
	30kW – three phase connection
Duration of scheme	20 years
Calculation methodology	To be confirmed

South Australia

South Australia (SA) was the first state in Australia to implement a FIT scheme. It was legislated through the Electricity (Feed-in Scheme – Solar Systems) Amendment Act 2008 which commenced in July 2008 (see Table 6.12). The net FIT was paid at a rate that was set to be double the retail price of electricity at AU$0.44/kWh. The systems are capped at different sizes according to the connection type (i.e. 10kW for single phase connection and 30kW for three phase connection).

Table 6.12 *South Australia feed-in tariff details*

Date of first Implementation	July 2008
Types of eligible renewables	Solar
Sectors eligible	Residential, small business and other facilities
Purchase obligation	AU$0.44/kWh
Scheme type	Net
Generation cap	160MWh per year
Capacity cap	10kW – single phase connection
	30kW – three phase connection
Duration of scheme	20 years
Calculation methodology	To be confirmed

Tasmania

In March 2008 it was announced that the Tasmanian government was considering the implementation of a mandatory FIT (see Table 6.13). Submissions were invited and accepted up to November 2008 to obtain input for the development of policy around a net FIT. This FIT is for solar and other renewable sources of electricity which are generated by households and small businesses. While awaiting further announcements regarding the implementation of FIT, one major retailer is currently offering FIT to PV installations at a rate of AU$0.2/kWh.

Table 6.13 *Tasmania feed-in tariff details*

Year of first implementation	Retail offering available
Types of eligible renewables	Solar
Sectors eligible	To be confirmed (TBC)
Purchase obligation	AU$0.2/kWh
Scheme type	Net
Generation cap	TBC
Capacity cap	TBC
Duration of scheme	TBC
Calculation methodology	TBC
Features	TBC

Victoria

The Victorian government has proposed a net FIT through the Energy Legislation Amendment Act 2007, which is expected to come into effect in 2009. Victoria is planning for two types of FITs to be in place (premium and standard) for renewables that are of different scales (see Tables 6.14 and 6.15). It was announced in May 2008 that the premium FIT will allow households to be paid AU$0.6/kWh for the excess electricity exported into the grid from the PV system. The standard

FIT for biomass, hydro, solar and wind commenced in January 2008 and is based on the 'fair and reasonable' criteria for the FIT rate. This ensures that generators are offered a 'fair price' by retailers, which is usually the equivalent retail electricity price, factoring in any additional costs that may also be incurred.

Table 6.14 *Victoria premium feed-in tariff details*

Year of first implementation	Announced but not legislated
Types of eligible renewables	Solar
Sectors eligible	Domestic
Purchase obligation	AU$0.6/kWh
Scheme type	Net
Generation cap	TBC
Capacity cap	3.2kW
Duration of scheme	15 years
Calculation methodology	To be confirmed (TBC)
Features	TBC

Table 6.15 *Victoria standard feed-in tariff details*

Date of first implementation	January 2008
Types of eligible renewables	Biomass, hydro, solar and wind
Sectors eligible	To be confirmed (TBC)
Purchase obligation	Equivalent to electricity prices
Scheme type	Net
Generation cap	TBC
Capacity cap	100kW
Duration of scheme	Not applicable
Calculation methodology	Retail prices

Western Australia

The Western Australia government has committed to the implementation of a gross FIT scheme which pays a rate of AU$0.6/kWh for all generation, which is even higher than that of the ACT (see Table 6.16). The Sustainable Energy Development Office within the Office of Energy which is developing the scheme also states that the FIT will be paid until the systems costs are covered, taking into account the capital subsidies, grants and rebates. After this, the generators will revert to the Renewable Energy Buyback Scheme. At the time of writing, they are still determining a number of design features including the commencement date of the scheme and the eligibility of systems.

Table 6.16 *Western Australia feed-in tariff details*

Year of first implementation	To be confirmed (TBC)
Types of eligible renewables	Solar
Sectors eligible	Domestic
Purchase obligation	AU$0.6/kWh
Scheme type	Gross
Generation cap	TBC
Capacity cap	10kW
Duration of scheme	20 years
Calculation methodology	TBC
Features	TBC

6.4. AFRICA

As described in Chapter 5, FITs have the flexibility to be adapted to the special framework conditions of developing countries and emerging technologies. Besides the implementation of FITs in countries like Argentina, Brazil, China, Israel and Pakistan, several African countries have shown increasing interest. The spark was lit by Mauritius, which has been successfully operating a FIT scheme for many years in order to incentivize cogeneration using bagasse, a by-product of the sugar industry. Kenya and South Africa, however, were the first countries to implement a full-scale FIT scheme covering a large number of renewable energy technologies. We will present both approaches below. In addition, several other countries, including Nigeria and Ghana, are currently working hard to establish well-designed FIT schemes.

6.4.1 Kenya

Kenya was the first African country to implement a FIT scheme based on policy transfer and international best practice. Traditionally, Kenya – like many other African countries – relies primarily on large-scale hydropower projects for electricity generation. At the moment, about 60 per cent of the national electricity demand is covered by this source. However, the increasing power demand and the decreasing precipitation due to climate change is forcing the national government to diversify its energy supply. The ideal geographic location of the country enables the government to count on another renewable energy source. The Great Rift Valley, stretching all the way from Syria to Central Mozambique, offers optimal conditions for geothermal power plants. This renewable energy technology already accounts for 10 per cent of the national power generation and is the cheapest of all technologies, including fossil-fuel-based power plants, which account for the remaining 30 per cent of power generation. On average, electricity demand has grown by 8 per cent annually.

While large hydro and geothermal power plants are already competitive without any support, the government decided to implement a FIT scheme in order to profit from other renewable energy technologies as well. According to the legislation which came into force in March 2008, biomass, small-scale hydropower and wind energy are eligible under the scheme (Kenyan Ministry of Energy, 2008). For each technology, a specific tariff was established that mirrors generation costs. Moreover, the tariff is differentiated for 'firm' and 'non-firm' power generation in the case of biomass and hydropower. 'Firm' refers to steadily produced electricity in line with a certain forecast and 'non-firm' refers to fluctuating electricity generation. As the producer of wind power has no means of controlling the power output, this differentiation does not apply for wind power. Besides, the tariffs for hydropower vary between US$0.06 and 0.12/kWh depending on the size of the plant. The duration of tariff payment is fixed at 15 years. The Kenyan FIT is financed by distributing the costs among all electricity consumers. The law determines that the grid operator can add all costs higher than US$0.026/kWh to the pass-through costs.

The Kenyan FIT scheme includes both a cap for the total installed capacity and one for the installed capacity of each individual plant, in order to control the costs for the final consumer and to integrate renewable energies in national planning (for capacity caps also see Sections 4.8 and 5.1). Tariff payment is guaranteed for wind power plants up to 150MW, biomass up to 200MW and hydropower up to 500MW. At the same time, the Kenyan legislator limited the maximum size of each power plant to 50MW in the case of wind power, 40MW for biomass and 10MW for small-scale hydro (see Table 6.17).

To reveal the economic competitiveness of renewable energy sources with fossil fuels, the Kenyan Ministry of Energy published a table which depicts the cost of energy based on the use of diesel generators (see Table 6.18). The table shows that, depending on the global price of crude oil, today renewable energy sources promoted by the FIT scheme are already in the same price range or cheaper.[5] Moreover, the price for renewable energies under the FIT remains stable or decreases, while the year 2008 has shown that global crude oil price can go far beyond US$70 per barrel.

Table 6.17 *Tariff payment under the Kenyan feed-in tariff scheme*

Technology	Tariff	Size of power plant
Wind	US$0.09/kWh	<50MW
Biomass (firm)	US$0.07/kWh	<40MW
Biomass (non-firm)	US$0.045/kWh	<40MW
Hydro (firm)	US$0.08–0.12/kWh	500kW–10MW
Hydro (non-firm)	US$0.06–0.1/kWh	500kW–10MW

Source: Jacobs, 2009

Table 6.18 *The energy cost associated with using medium-speed diesel plants in Mombasa and Nairobi (August 2007)*

Global crude oil price (US$/bbl)	Fuel cost (US$/kWh)	
	Mombasa	Nairobi or upcountry
50	0.067	0.079
60	0.078	0.09
70	0.089	0.101

Source: Kenyan Ministry of Energy, 2008

One year after the implementation of the Kenyan FIT scheme the responsible ministry seemed satisfied with the industry's response. Several power producers have expressed interest in realizing renewable energy projects in Kenya and six producers have already undertaken site-specific feasibility studies for wind power projects. In total, all intended projects have an installed capacity of 500MW (Ondari, 2009). However, the law has been criticized for allowing exemptions from the purchase obligation and for establishing maximum tariffs instead of minimum tariffs (see Chapter 4, Bad FIT Design).

6.4.2 South Africa

South Africa is an interesting case study as the approval of the FIT scheme was related to the end of a long debate on nuclear power. The country, which covers almost 90 per cent of electricity demand with national coal reserves, intended to diversify the national energy mix by building at least two nuclear power plants. Due to a severe energy crisis and frequent power outages the government urgently needed additional power generation capacity.

In December 2008 the South African government abandoned its plans to build several nuclear power plants. The country wanted to deploy 20,000MW of installed nuclear capacity by 2025. The total costs for these projects were expected to be $12 billion. Two companies were involved in the bidding process: the French company Areva, which proposed to build two 1650MW reactors, while Westinghouse from the US planned three 1140MW reactors (Derby and Lourens, 2008). The decision to scrap the ambitious nuclear energy plans was largely due to the high costs. The ongoing global financial and economic crisis further increased the cost of capital for such large-scale investment.

Now, South Africa wants to meet future energy demand by significantly increasing the share of renewable energies. In the same month the government decided to scrap its nuclear plans, the National Energy Regulator of South Africa (NERSA) issued a consultation paper for an ambitious FIT scheme (NERSA, 2008). According to the national renewable energy plans, by 2013 about 10,000MWh of green electricity will be produced every year. In addition, energy efficiency measures should account for a saving of 3000MWh by 2012 and a further 5000MWh by 2025 (EIA, 2008).

The FIT proposal of December 2008 was followed by an important consultation process before the final piece of legislation was issued in March 2009 (NERSA, 2009). This process included a public hearing in February 2009, and important changes were made during the consultation process. As shown in Table 6.19, the tariffs for all technologies were significantly increased as NERSA had based its calculations on outdated information. In the case of wind power, it initially proposed a tariff of R0.75/kWh (about €0.05.8/kWh). Finally, a tariff of R1.25 (€0.104/kWh) was granted for wind power producers. Interestingly, South Africa also promotes concentrated solar power (CSP) and therefore does not exclusively focus on the least-cost renewable energy technologies. Furthermore, landfill gas and small-scale hydropower (less than 10MW) will be eligible under the support mechanism. Biomass pulp and paper and bagasse were excluded from the FIT scheme as those technologies were already supported under different programmes. The legislator left the door open for including more renewable energy technologies in the near future.

The South African FIT scheme guarantees tariff payments for a period of 20 years. Afterwards, the producers will have the possibility of bilaterally negotiating power purchasing agreements with the grid operator. In line with international best practice, the tariffs are based on the generation costs for each technology. The South African tariff calculation methodology is further described in Section 2.3.3.

The initial intention to automatically reduce tariff payment every year (see Section 3.9) was scrapped during the consultation process as most stakeholders argued that this kind of provision is not necessary in the first years of operation. For new installations, the tariffs will be indexed to inflation every year (see Section 3.12). The feed-in scheme also includes a periodic revision of the support mechanism. In the first five years of operation, the FIT is going to be reviewed every year and thereafter every three years (see Section 2.13).

Table 6.19 *Tariff payment under the South African feed-in tariff scheme*

Technology	First tariff proposal (2008) in € (Rand in brackets)	Tariffs as approved in 2009 in € (Rand in brackets)
Landfill gas	0.033/kWh (0.4321)	0.075/kWh (0.9)
Small hydro (less than 10MW)	0.057/kWh (0.7376)	0.078/kWh (0.94)
Wind power	0.051/kWh (0.6548)	0.104/kWh (1.25)
Concentrating solar power (CSP)	0.047/kWh (0.6064)	0.175/kWh (2.10)

Source: Jacobs, 2009

6.5 Asia

FITs in Asia are a mixed bag. Research on India produced various opinions on whether they have 'true' FITs or not. China has something FIT-like for biomass, has now introduced a FIT for wind and is expected to establish a FIT for utility-scale solar plants by the end of 2009. Japan has seen efforts to push renewable energy forward against opposition from the monopoly energy companies. The present national renewable electricity deployment target stands at a far from spectacular 1.63 per cent by 2014 (compare with EU targets of around 20 per cent by 2020). Solar PV FITs may emerge in Japan first. The Greater Tokyo area is working on the implementation of a FIT scheme. South Korea has a FIT; a large part of the 2009 economic stimulus package is devoted to RE (see Chapter 1) and the desire to develop a major manufacturing industry. Malaysia, Singapore and Taiwan are all exploring the implementation of a FIT.

6.5.1 India

India has become the third fastest-growing wind power nation, behind China and the US. In 2008 they installed 1.8GW, and are fifth in overall renewables capacity worldwide, including large hydro (REN21, 2009, p11). The country is seeing the introduction of new national and state-level policies aimed at fostering a PV manufacturing industry. Capital investment subsidies of 20 per cent are a key incentive, as well as accelerated tax depreciation and the establishment of a national FIT for grid-connected PV (REN21, 2009, p15). The programme was initially capped at 50MW through to 2009, but this may rise to 1000MW in a second phase. The tariff pays up to INR12/kWh (more details below) (REN21, 2009, p31).

There have been some other relevant initiatives aimed at reforming the energy sector. They include the Electricity Act 2003 legislated by the Ministry of Power, which contains provisions for the promotion of renewable energy. Sections 86(1)(e) and 61(h) are clear developmental provisions relating to renewables, called Renewable Purchase Obligations (RPOs). While section 61(h) is important from the perspective of increasing the viability of the project, section 86(1)(e) helps develop the market for renewable energy projects. It allows the regulator to set out a minimum percentage of renewable energy which must be purchased by distribution companies, but the provisions are inadequate in that they are only enabling and not mandatory.

Although the Electricity Act 2003 mandates for all the states in India to implement such purchase obligations, by May 2009 only 14 State Electricity Regulatory Commissions (SERCs) out of the 28 states in India have issued orders or regulations under the RPO. According to these specifications, the distribution licensees are required to purchase specified amounts of renewable electricity out of their total electricity portfolios. These obligations increase annually from 0.5 to 10 per cent over a period of 10–20 years depending on decisions made within the individual states.

In India, the importance of the role of renewable energy in the transition to a sustainable energy base was recognized as early as the 1970s. At the Government level, political commitment to renewable energy manifested itself in the establishment of the Department of Non-Conventional Energy Sources in 1982, which was subsequently upgraded in 1992 to a full-fledged Ministry of Non-Conventional Energy Sources, renamed Ministry of New and Renewable Energy (MNRE) in October 2006.[6]

India is set to formulate a Renewable Energy Act with a national target to meet 20 per cent by 2020. At present the share is 9 per cent. The central government, in a bid to give a major push to the renewable sector, has appointed a high-level committee to prepare the national law. It may have a FIT as a policy instrument, and include measures for strict implementation of its purchase agreement and penalties for breach of such agreements. Certain state electricity regulatory commissions have already passed orders with regard to renewables purchases, but these need to be spread across the country to make sure that all the states adopt the policy. Similarly, the law should enable the imposition of penalties for violation of such purchases; at present some state regulatory commissions have passed orders in this regard.

India's installed wind power capacity is growing fast, and stands at over 9650MW (REN21, 2009, p23). The MNRE introduced a generation-based incentive scheme for grid-connected wind power projects in July 2008 wherein eligible producers can receive a payment of INR0.50/kWh. An approved project will receive this for ten years. Tamil Nadu, Karnataka, Gujarat and Maharashtra are some of the leading wind energy states in India and all these states have purchase obligations.

Grid-connected solar power in India currently amounts to only 2.12MW. With a view to developing and demonstrating the technical performance of grid-connected solar power projects, the MNRE has established guidelines for generation-based incentives for grid-connected solar power generating projects for the first time in January 2008. The notional cap was fixed at INR15/kWh for solar PV and INR13/kWh for solar thermal grid-connected projects. According to the guidelines, the utility will purchase the generated power at the highest rate of traded power from other sources. MNRE would provide funds for the deficit, to a maximum of INR12/kWh for solar PV and INR10/kWh for solar thermal. This will be valid for projects commissioned up to December 2009; thereafter the amount will be reduced by 5 per cent. The generation-based incentive will be available for a period of ten years from the date of commissioning.

West Bengal was the first state in the country to provide a tariff for solar power. Rajasthan and Gujarat in western India and Punjab in northern India are some of the other leading states with a solar FIT. All the states have huge potential for harnessing solar power because of their peculiar geographical location and they have big investments lined up for solar energy production.

Apart from various political parties devoting some space for green issues in their political manifestos, in terms of increase in the percentage of renewable energy to be sourced, climate change was hardly a poll issue in the 2009 parliamentary elections. Environmental NGOs hope that whichever coalition forms the next national government, they will introduce a strong and effective national law.

REFERENCES

BGB (2008) 'Gesetz zur Neuregelung des Rechts der erneuerbaren Energien im Strombereich und zur Änderung damit zusammenhängender Vorschriften', *Bundesgesetzblat*, vol 1, no 49, p2074

BMU (2007) *Background Information on the EEG Progress Report 2007*, BMU, Berlin, www.bmu.de/files/pdfs/allgemein/application/pdf/eeg_kosten_nutzen_hintergrund_en.pdf

BMU (2008) *Entwicklung der Erneuerbaren Energien in Deutschland, Stand: 12 Daten des Bundesministeriums zur Entwicklung der Erneuerbaren Energien in Deutschland im Jahr 2007 (vorläufige Zahlen) auf der Grundlage der Angaben der Arbeitsgruppe Erneuerbare Energien-Statistik*, AGEE-Stat, Berlin

Bode, S. (2006) *On the Impact of Renewable Energy Support Schemes on Power Prices*, Research Paper, Hamburg Institute of International Economics (HWWI), Hamburg

BOE (2007) 'Real Decreto 661/2007, de 25 de mayo, por el que se regula la actividad de producción de energía eléctrica en régimen especial', *Boletín Oficial del Estado*, no 126, 26 May, p22846, www.boe.es/boe/dias/2007/05/26/pdfs/A22846-22886.pdf

BOE (2008) 'Real Decreto 1578/2008, de 26 de septiembre, de retribución de la actividad de producción de energía eléctrica mediante tecnología solar fotovoltaica para instalaciones posteriores a la fecha límite de mantenimiento de la retribución del Real Decreto 661/2007, de 25 de mayo, para dicha tecnología', *Boletín Oficial del Estado*, no 234 of 27 November, p39117, www.boe.es/boe/dias/2008/09/27/pdfs/A39117-39125.pdf

Cory, K., Couture, T. and Kreycik, C. (2009) *Feed-In Tariff Policy: Design, Implementation, and RPS Policy Interactions*, National Renewable Energy Laboratory, NREL/TP-6A2-45549, Golden, CO

Couture, T. (2009) *State Clean Energy Policy Energy Policy Analysis: Renewable Energy Feed-in Tariffs*, National Renewable Energy Laboratory, Golden, CO

de Miera, S., del Río, G. and Vizcaíno, I. (2008) 'Analysing the impact of renewable electricity support schemes on power prices: The case of wind electricity in Spain', *Energy Policy*, vol 36, no 9, pp3345–3359

Derby, R. and Lourens, C. (2008) 'South Africa scraps plan to build nuclear power plant', www.bloomberg.com/apps/news?pid=20601085&sid=a2kESlbhYYHE&refer=europe

Dinica, V. and Bechberger, M. (2005) 'Spain – Country Report', in Danyel Reiche (ed.) *Handbook of Renewable Energies in the European Union*, Peter Lang, Berlin, pp263–279

Edison Electric Institute (1996) *Statistical Yearbook of the Electric Utility Industry*, EEI, Washington, DC

EIA (2008) *South Africa, Country Analysis Briefs*, Energy Information Administration, October 2008, www.eia.doe.gov/cabs/South_Africa/pdf.pdf

Farrell, J. (2009) 'Feed-in tariffs in America: Driving the economy with renewable energy policy that works', *New Rules Project*, www.newrules.org/energy/publications/feedin-tariffs-america-driving-economy-renewable-energy-policy-works

Gipe, P. (1995) *Wind Energy Comes of Age*, John Wiley & Sons, New York, NY

Gipe, P. (2006) *Renewable Energy Policy Mechanisms*, Wind Works Organization, Tehachapi, CA, www.wind-works.org/FeedLaws/RenewableEnergyPolicyMechanisms byPaulGipe.pdf

Gipe, P. (2009) *Vermont FITs Become Law: The Mouse that Roared, First North American Jurisdiction with Small Wind Tariff*, May 28, 2009, www.wind-works.org/FeedLaws/USA/VermontFITsBecomeLawTheMouseThatRoared.html

Girardet, H. and Mendonça, M. (2009) *A Renewable World: Policies, Practices and Technologies*, Green Books, Totnes

Grace, R., Rickerson, W., Porter, K., DeCesaro, J., Corfee, K., Wingate, M. and Lesser, J. (2008) *Exploring Feed-In Tariffs for California: Feed-In Tariff Design and Implementation Issues and Options*, CEC-300-2008-003-D, California Energy Commission and Kema Consulting, Sacramento, CA

Grace, R., Rickerson, W., Corfee, K., Porter, K. and Cleijne H. (2009) *California Feed-In Tariff Design and Policy Options*, CEC-300-2008-009F, California Energy Commission, Oakland, CA

Green Energy Act Alliance (2009) *Ontario Gives Green Energy Act the Green Light: A first for North America - Ontario's Green Energy and Economy Act Becomes Law*, 14 May, www.greenenergyact.ca/Page.asp?PageID=924&ContentID=1259

Held, A. (2008) *RES-e Support in Europe – Status Quo*, Presentation at the Futures-e final conference, Brussels, 25 November, www.futures-e.org/Final_Conference/RES-E_Overview_futures-e_Brussels_(Held,%2025112008).pdf

IfnE (2009) *Beschaffungsmehrkosten der Stromlieferanten durch das Erneuerbare-Energien-Gesetz 2008 (Differenzkosten nach § 15 EEG)*, Ingenieurbüro für Neue Energie (IfnE), Gutachten im Auftrag des Bundesministeriums für Umwelt, Naturschutz und Reaktorsicherheit, Teltow, March

IfnE/BMU (2007) *Ökonomische Wirkung des Erneuerbare-Energien-Gesetzes, Zusammenstellung der Kosten- und Nutzenwirkung*, Ingenieurbüro für Neue Energie (IfnE), Untersuchung im Auftrag des Bundesministeriums für Umwelt, Naturschutz und Reaktorsicherheit, 30. November

Jacobs, D. (2008) *Analyse des spanischen Fördermodells für Regenerativstrom unter besonderer Berücksichtigung der Windenergie*, BWE Research Paper, September 2008.

Jacobs, D. (2009) *Renewable Energy Toolkit – Promotion Strategies in Africa*, World Future Council, April

Kenyan Ministry of Energy (2008) *Feed-In Tariff Policy on Wind, Biomass, and Small-Hydro Resource Generated Electricity*, March, www.investmentkenya.com/index.php?option=com_docman&Itemid=&task=doc_download&gid=20

Lean, G. (2009) 'We are one step closer to clean coal', *The Independent*, 26 April, www.independent.co.uk/opinion/commentators/geoffrey-lean-we-are-one-step-closer-to-clean-coal-1674258.html

Martin, T. (2009) *Solar Photovoltaic Installation Potential under a UK Feed-in Tariff*, University of Bristol School of Chemistry, available from author

Martinot, E., Wiser, R. and Harmin, J. (2006) *Renewable energy policies and markets in the United States*, Center for Resource Solutions www.resource-solutions.org/lib/librarypdfs/IntPolicy-RE.policies.markets.US.pdf, accessed 20 August 2008

Morris, C. (2009) 'FITs in the USA', *PV Magazine,* March, pp20–25

NERSA (2008) *NERSA Consultation Paper – Renewable Energy Feed-in Tariff,* National Energy Regulator of South Africa, December 2008

NERSA (2009) *South African Renewable Energy Feed-in Tariff (REFIT), Regulatory Guidelines*, National Energy Regulator of South Africa, 26 March

Nitsch, J. (2008) *Lead Study 2008, Further Development of the 'Strategy to Increase the Use of Renewable Energies' Within the Context of the Current Climate Protection Goals of Germany and Europe*, German Federal Ministry for the Environment, Nature Conservation and Nuclear Safety (BMU), October, Stuttgart

Ondari, J. (2009) Renewable energy policy spurs investor interest', *Daily Nation*, 16 March, www.nation.co.ke/business/news/-/1006/538268/-/view/printVersion/-/13tgqkf/-/index.html

Ontario Power Authority (2009) *Renewable Energy Supply Survey Results*, February

REN21 (2009) *Renewables Global Status Report: 2009 Update*, REN21 Secretariat, Paris, www.ren21.net/pdf/RE_GSR_2009_update.pdf

Rickerson, W. and Grace, R. C. (2007) *The Debate over Fixed Price Incentives for Renewable Electricity in Europe and the United States: Fallout and Future Directions*, Heinrich Boll Foundation, Washington, DC

Seager, A. and Gow, G. (2007) 'Britain accused of scuppering EU's renewable energy plan', *The Guardian*, 13 October, www.guardian.co.uk/business/2007/oct/13/europeanunion.renewableenergy

Sensfuß, F., Ragwitz, M. and Genoese, M. (2008) 'The merit-order effect: A detailed analysis of the price effect of renewable electricity generation on spot market prices in Germany', *Energy Policy*, vol 36, no 8, pp3076–3084

Sösemann, F./BMU (2007) *EEG: The Renewable Energy Sources Act: The Success Story of Sustainable Policies for Germany*, Federal Ministry for the Environment, Nature Conservation and Nuclear Safety, Berlin, Germany

Sovacool, B. K. (2008) *The Dirty Energy Dilemma: What's Blocking Clean Power in the United States*, Praeger, Greenport, CT, p233–234

Wenzel, B. and Nitsch, J. (2008) *Ausbau Erneuerbarer Energien im Strombereich, EEG-Vergütungen, Differenzkosten und – Umlagen Sowie Ausgewählte Nutzeneffekte bis 2030*, Teltow, Stuttgart, December

7

Dispelling the Myths about Technical Issues

Many energy providers, electric utilities, politicians and even ordinary citizens believe that renewable power plants cannot provide reliable power because the electricity generated from the fuels they rely upon, such as wind, water and sunlight, is variable. The director of one prestigious research institute in the US went so far as to tell one of the authors that forcing renewables to operate as reliably as conventional fossil fuel or nuclear units – trying to devise renewable systems that could overcome their variability – was like 'trying to make a pig fly: you won't succeed and you only make the pig unhappier'.[1] FITs, in essence, are viewed by some as being as ridiculous as trying to give pigs wings to make them fly.

Interconnecting small-scale distributed renewables is deemed even more problematic, as having thousands of decentralized plants (compared to a handful of centralized facilities) is believed to complicate grid management, perhaps insurmountably. People who have this view can be compared with those in the 1980s who never managed to believe that something like the futuristic 'internet' would ever work, or those at the turn of the 20th century who thought humankind would never be able to fly. Other stakeholders think that because some renewable energy technologies are variable, using them requires backup capacity from coal and gas power plants, meaning FITs force people to pay twice, once for the renewable plant and again for the backup. One Vice President of a large US utility stated that it would cost six times as much in backup costs to deploy wind compared to coal in his service territory.[2] Others say that grid integration is technically manageable, but that the costs of transmission and integration are too expensive. Therefore, many utilities and power operators consider renewable power plants non-dispatchable, incapable of providing 'base-load' power (electricity available all the time), and hence inferior.

This chapter shows that such conventional thinking is completely wrong. Virtually all renewable systems already operate *more* reliably than conventional units. Some renewable power sources such as geothermal, biomass, solar thermal and hydroelectric units already provide continuous and reliable power around the clock, on demand. The variability of resources such as wind and solar can be smoothed out through shrewd planning and storage technologies and through backup capacity from the above-mentioned technologies. Solar power is excellent at displacing peak demand and displacing retail power needs. Dozens of scientific studies based on real-world experience show that the costs of grid integration and constructing transmission lines are negligible. Far from becoming an insurmountable technical barrier, this chapter shows that technical issues relating to intermittency, interconnection and transmission have largely been resolved by engineers and system operators. What remains as an important task is advancing the knowledge of these solutions to the rest of us.

7.1 THE UNRELIABILITY OF CONVENTIONAL UNITS

To penalize renewables for their variability or intermittency not only ignores how that variability can be mitigated, it also obscures equal amounts of variability inherent in conventional fossil fuel and nuclear resources. All electricity systems must respond to the complex interplay of constantly changing supply and demand. They are subject to unexpected failures and outages and influenced by a large number of planned and unplanned events. Daily load variances occur, as routine practices reinforce the effects of changing from day to night, such as turning lights on, raising indoor temperature when waking up, taking showers before breakfast, cooking in the dinner hour and washing dishes, or charging electric vehicles at night. Over the course of a week, energy use changes as the weekend approaches and, throughout the year, as seasonal differences in temperature and climate occur. While it is certainly true that the output from conventional power plants can be measured quite accurately, researchers from the Lawrence Berkeley National Laboratory and the American Council for an Energy-Efficient Economy noted that virtually 'every other aspect of planning for and implementing that resource is riddled with uncertainty' (Vine et al, 2007).

Four types of uncertainty are most common: unexpected outages, variance in construction costs, variance in demand forecasts, and transmission and distribution vulnerability. And, perhaps surprisingly, renewable power plants address *each* of these types of variability better than conventional units:

1   Let us begin by discussing the unplanned outages for conventional units. The average coal plant operating on the market today is out of service 10–15 per cent of the time (Sovacool, 2009). Looking at the performance of conventional generators in the US from 2000 to 2004, the North American Electric Reliability Corporation found that plants shut down for scheduled maintenance 6.5 per cent of the year and require unscheduled maintenance or experience forced outages another 6 per cent of the year. Their study noted that conventional output is guaranteed on average only 87.5 per cent of the

time in the US, with a range of 79–92 per cent (NERC, 2005). To cope with the variability of conventional units, system operators must operate a 15 per cent reserve margin of extra capacity, much of which is continually fuelled and spinning ready for instant use.

Nuclear plants are not much better. One survey of nuclear power plant operating performance for US, French, Belgian, German, Swedish and Swiss reactors found mean durations of continual operation from 35 to 88 days (Perin, 1998). In other words, the average plant only operated one to three months without some sort of unplanned outage event, half of which were related to equipment failure. Of all 132 nuclear power plants built in the US (only 52 per cent of the 253 originally planned), almost one-quarter (21 per cent) were permanently and prematurely closed due to reliability or cost problems, and 27 more have failed for a year or more at least once (Lovins et al, 2008). Even reliably operating nuclear plants must shut down 39 days every 17 months for refuelling and scheduled maintenance. They must also shut down during blackouts, and then take incredibly long times to restart. During the August 2003 blackout in the US, nine perfectly operating nuclear plants had to shut down and then took 12 days to restart. During the first three days, when they were most needed, their output was below 3 per cent (Lovins et al, 2008). Regions heavily dependent on a fleet of nuclear plants are at greater risk because drought or safety problems can close many units simultaneously.

2 Conventional plants are more prone to cost overruns and manufacturing glitches. These power plants are 'lumpy systems' in the sense that additions are made in large 'lumps' (such as 1000MW reactors). These facilities have long lead times, making them vulnerable to project delays, unforeseen events, cost overruns and project cancellations. Nuclear power plants in Canada, the US and Finland are a prime example here. In Canada, delays and cost overruns on nuclear power plants accounted for CA$15 billion of 'stranded debt' created by Ontario Hydro (Winfield et al, 2006). In the US, the actual construction cost for 75 nuclear power plants was quoted to be US$89.1 billion, but because of project delays and manufacturing errors, cost overruns ballooned to more than *three times* as much, at US$283.8 billion (US Congressional Budget Office, 2008). The Finnish nuclear power plant at Olkiluoto was expected to cost €3 billion. By now the costs have risen to at least €4.5 billion and the power plant which was to be completed by 2009 will not go online before mid-2012.

3 Gargantuan conventional plants, because they take longer to build, are also at greater risk of unexpected changes in electricity demand over long periods of time. We have a hard enough time predicting the weather or the outcome of political elections; imagine the difficulty of projecting how an entire sector will demand electricity five, ten, or even twenty years from now. In the 1970s and 1980s, excessively high forecasts of growth in demand for electricity led to overbuilding of generating plants and massive electric system cost overruns in many states. One infamous example was in Washington State,

where the Washington Public Power Supply spent more than $5 billion partially constructing nuclear plants that were later abandoned when demand for electricity dropped. Between 1972 and 1984, more than $20 billion in construction payments flowed into 115 nuclear power plants worldwide that were subsequently abandoned by their sponsors because they were no longer needed (Cavanagh, 1986).

4   Both sets of large plants, fossil fuelled and nuclear, must rely on brittle transmission lines easily disrupted by lightning strikes, storms, squirrels and bullets. Given that more than 98 per cent of blackouts and power outages start on the grid, such centralization has grave risks for electricity reliability (Lovins et al, 2008).

The renewable resources supported by FITs, ironically, respond better to each of these problems. Modern wind turbines and solar panels have a technical reliability above 97 per cent. Such high reliability is for one wind turbine or solar panel, so any amount of significant wind or solar power in an electricity system would never see all (hundreds of thousands of units) down at the same time. When individual units do rarely fail, they do so in smaller increments. The high technical reliability for wind and solar lowers the probability of unplanned outages and lessens the need for operational and capacity reserves (Jacobson and Masters, 2001). Since forced outages for conventional units range from 10 to 15 per cent, and the wind turbine failure rate is less than 3 per cent, the extent that wind replaces fossil fuels improves the reliability of the system by 7–12 per cent (and also reduces backup requirements by an equivalent amount). New inverter technology has the potential to enhance the reliability of solar even further, as it will enable systems to work when partially shaded.

In terms of modularity, construction cost overruns, and rapid alterations in electricity demand, the quicker lead times for renewable power plants and small-scale units enables a more accurate response to load growth or reduction. Wind farms, geothermal power plants, and biomass plants often take between one and two years to construct, and if the units are available, solar panels can be installed in as little as a few months. Small-scale solar and wind units can be matched to serve almost any load, and medium- to commercial-scale wind turbines, bioelectric plants and geothermal stations can be installed in increments ranging from 1.5MW to 20MW. Such modularity minimizes the financial risk associated with borrowing hundreds of millions of dollars to finance plants for ten or more years before they start producing a single kWh of electricity, and it means electricity loads can be precisely matched.

Finally, in terms of transmission and distribution vulnerability, the small-scale and distributed renewable power generators promoted by FITs can improve grid reliability, lessen the need to build expensive transmission infrastructure, reduce congestion, offer important ancillary services, and improve energy security through geographic diversification. Deploying distributed solar, biomass and small-scale wind units offers an effective alternative to constructing new transmission and distribution lines, transformers, local taps, feeders and switchgears, especially in congested areas or regions where the permitting of new transmission networks is difficult. The Pacific Gas and Electric Company, the largest investor-owned utility in California, built an entire power plant in 1993 to test the grid benefits of a 500kW distributed solar power plant. The utility found that the distributed solar plant improved voltage support, minimized power losses, lowered operating temperatures for transformers on the grid, and improved transmission capacity. The benefits were so large that the small-scale generator was twice as valuable as estimated, with projected benefits of $0.14–0.2/kWh (Wenger et al, 1994). This could be why the Institute of Electrical and Electronics Engineers in the US recently concluded that dispersed renewable resources such as wind can be managed not only through interconnection and integration without degrading the network, they can also contribute to improvements in system performance (Smith et al, 2007).

7.2 The Reliability of Hydro, Geothermal, Solar Thermal and Biomass

Commercial hydroelectric, geothermal, bioelectric and biogas power plants provide predictable, 24-hour base-load power in many parts of the world, including the US (where they satisfy more than 7 per cent of national electricity demand). Other countries, like Norway, rely entirely on these technologies. Equally, the latest solar thermal power plants can now provide reliable electricity as they operate in combination with molten salt and other large storage units.

These power facilities provide reliable power without the need for backup. Many of these systems are subject to woeful underinvestment, yet both hydropower and geothermal plants could provide almost the entire world's electricity *by themselves* if their technical potential was fully tapped. The world consumed about 17,000TWh of electricity in 2007, yet a comprehensive study undertaken by the International Energy Agency and others identified 14,370TWh of achievable remaining potential for hydroelectric facilities (International Hydropower Association, 2000) Similarly, the International Geothermal Association surveyed a collection of studies and concluded that 22,400TWh of geothermal power potential existed (Bertani, 2002).

It is always good to remember that when we are talking about the types of technologies that FITs promote, we are not talking only about intermittent resources such as wind and solar PV. We are also talking about big and small hydroelectric dams, solar thermal and geothermal plants, and bioelectric stations (some combusting fuel and others harvesting methane from landfills) that have been proven through decades of experience to operate identically to coal, oil, natural gas and nuclear units.

7.3 THE RELIABILITY OF INTERCONNECTED WIND AND SOLAR

While wind and solar systems are more variable than their hydro, geothermal, solar thermal and biomass counterparts, interconnecting dispersed wind and solar units greatly improves their reliability. Electrical and power systems engineers have long held the principle that the larger a system becomes, the less reserve capacity it needs. Demand variations between individual consumers are mitigated by grid interconnection in exactly this manner and modern communication technology enables us to make this happen. When a single electricity consumer starts drawing more electricity than the system has allocated for each consumer, the strain on the system is insignificant because so many consumers are drawing from the grid that it is entirely likely another consumer will be drawing less to make up the difference (International Energy Agency, 2005). This 'averaging' works in a similar fashion on the supply side of the grid. Individual wind turbines and solar panels average each other out in electricity supply. When the wind is not blowing through one wind farm or the sun not shining on someone's house, it is likely to be blowing harder or shining brighter near another. Therefore, the improvement of interconnection capacity between countries and regions is of special importance for renewable energy sources. Besides, modern, large-scale wind power plants are often remote-controlled by grid operators in order to increase or reduce electricity output according to demand (see Section 3.5).

A large number of meteorological wind studies make this point forcefully. Scientists looking at a 3-year data set for Scandinavian countries from 2000 to 2002 noted that that longest duration in low wind speeds per year was 58 hours for Denmark, 19 hours for Finland and Sweden, and 9 hours for Norway. However, none of these four rare events occurred at the same time, meaning there were no totally calm periods for all four countries together (Gul and Stenzel, 2006, p173). A separate study looking at Denmark and Germany found that the maximum hourly swing in wind speeds over a distributed network of wind farms rarely exceeded 20 per cent and had a standard deviation of hourly swings of 3 per cent. The study calculated that the maximum measured change in output per minute for a massive 2400MW wind farm would be less than 6MW, or 0.25 per cent of its total output (Gul and Stenzel, 2006, p171). Similarly, hourly wind data collected over a 23-year period from 66 different locations in the UK found that low wind speed events affecting more than half the country were very rare. For less than 10 per cent of the total time were wind speeds below 4 metres per second at individual sites, and there was no single event over the entire 23 years where wind speeds were low throughout all of the locations (Olz et al, 2007, p30). The conclusions advanced by these scientific studies are only bolstered by real-world operating experience in the US, Germany, Canada and the EU.

In the US, one study of utility experience with wind farms spread across locations in Minnesota, California, Wisconsin, New York, Oregon, Wyoming and Colorado found that greater penetration of wind plants *helped* grid operators handle major outages and contingencies elsewhere in the electricity network (DeMeo et al, 2005). Another assessment of 19 wind sites in the central US noted that almost all parameters from wind power improved as the number of interconnected sites increased, including standard deviations of array-average wind speed and wind power, reliability, and the need for energy storage or reserve capacity (Archer and Jacobson, 2007). A third study performed by General Electric for the Independent System Operator in New York investigated a 10 per cent wind penetration scenario in New York State, or the addition of about 3300MW of installed wind capacity on a 33,000MW peak-load system. When researchers posited that the wind capacity was located across 30 different sites, they found 'no credible single contingency' that led to a significant loss of generation. Because the system in New York was already designed to handle a loss of 1200MW due to the unreliability of conventional generators, it had more than enough resiliency to enable the incorporation of wind (Piwko et al, 2005). This could be why even though the US has more than 25,000MW of installed wind capacity (the largest absolute amount in the world), not a single conventional unit has been installed as a backup generator.

In Germany, the hundreds of thousands of dispersed solar photovoltaic units do not overwhelm system operators nor do they need highly advanced grids. Using a transmission and distribution system similar to the US, Germany integrates 350,000 separate solar installations (90 per cent of which are on residences) to provide 3.5GW of peak capacity. The highly dispersed and distributed nature of this resource means that when the sun shines in one area it often cancels out cloudiness in others, making it easier to manage. The German Solar Industry Association believes that solar penetration could be ramped up ten times to 35GW without any inherent technical problems.[3] Moreover, grid operators have proven that they can merely issue grid codes for the different voltage levels of the grid to increase network stability when needed.

In Canada, a study in Ontario investigated the impact of 20 per cent wind penetration on its electricity grid. The assessment accounted for seasonal wind and load patterns, daily wind and load patterns, changing capacity value for delivering power during peak load, and geographic diversity. It used wind and load data for one year and concluded that the more wind that existed in the system and the more geographically dispersed it was, the more it reduced volatility, in some cases by up to 70 per cent (AWS TrueWind, 2005).

Last, another study looked at the wind portfolios of all major power providers in the EU and found that a large contribution of wind was technically and economically feasible. The study noted that the more wind farms are interconnected, the more performance of wind turbines increases (and the costs of their electricity decreases). The study also found that extremely large shares of wind could be realized without compromising the security of the existing transmission and distribution system (European Wind Energy Association, 2005). When researchers followed up on their results with thousands of additional simulations in 2008 and 2009, they found that cross-border transmission of electricity from interconnected wind farms distributed across the EU would not negatively affect reliability. No single weather event or accident occurred that would affect wind farms in all or even most countries at the same time. Furthermore, they found that the effect of aggregating electricity from wind farms across multiple countries more than *doubled* the capacity factor of those interconnected wind turbines (Trade Wind, 2009).

These studies, in other words, conclusively show that widespread use of FITs would not compromise the stability of the electricity grid by incentivizing people to connect 'too many' renewables. The more FITs encourage the adoption of wind and solar, the more stable the grid becomes, rather than the other way around.

7.4 Hybrid Systems

Apart from connecting intermittent renewables of the same type, different renewables can be integrated together (or with energy efficiency measures and technologies) to create very reliable hybrid systems. Installing wind turbines at geothermal power plants creates effective base-load systems as wind data already exist at plant locations to site cooling towers, and plant designs allow for suitable spare land. These plants can rely on geothermal electricity to back up or offset any unexpected shortfalls in wind (Harvey, 2008). Similarly, wind farms can be coupled with biomass plants to completely eliminate their intermittency using agricultural wastes and residues, methane from landfills, energy crops, and trash as sources of fuel (Denholm, 2006).

A far more extensive hybrid system, called the 'combined power plant' and the 'renewable-energy combined cycle power station,' or *Kombikraftwerk*, exists in Germany. Operated by Schmack Biogas AG, SolarWorld AG and Enercon, this combined power plant relies on an integrated network of 36 wind, solar, biomass and hydropower installations spread across Germany. Wind and solar units generate electricity when those resources are available, and a collection of biomass and biogas plants and a pumped hydro facility make up the difference when they are not. The system can immediately adapt to a shortfall in any one resource by drawing on the others. As of early 2009, the 23.2MW combined power plant consisted of 11 wind turbines at three separate wind farms, 4 combined heat and power biogas units, 23 distributed solar systems and a pumped hydro storage plant linked via central control (see Figure 7.1). In 2008, the facility produced 41.1GWh of electricity without a single interruption of supply, enough electricity for 12,000 households in Schwäbisch Hall. This project shows quite clearly that a combination of different renewable energy technologies can potentially cover the entire electricity demand in Germany. The project size was chosen to represent the German electricity demand on a scale of one to ten thousand. The combined power plant also lowered

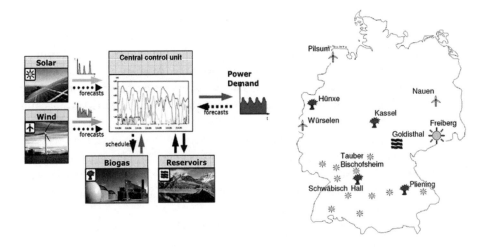

Figure 7.1 *The 23.2MW Kombikraftwerk, a hybrid wind–solar–biogas–hydro facility serving Schwäbisch Hall, Germany*

Source: Topfer, 2006

the region's dependence on oil and natural gas, and produced no greenhouse gas emissions. This combination of different renewable energy technologies can even be incentivized by special FIT design options (see Section 3.4).

A similar hybrid system exists in the Saxony-Anhalt district of Germany near the Harz district. There, 6MW of wind are connected to an 80MW pumped hydro facility used to back up wind output by pumping water up when the wind is available, and then using gravity to power two 40MW turbines to balance the system when the wind is not. In the village of Dardesheim, the wind–hydro system is in the process of being integrated with distributed solar power plants, six biogas systems, and a large 5MW cogeneration unit fuelled by recycled vegetable oil. The resulting wind–hydro–solar–biogas–vegetable oil facility, integrated via a digital control station, is expected to provide about 500 million kWh of electricity annually to a region that consumes only 800 million kWh, meaning it will meet two-thirds of all electricity demand (Federal Ministry of Economics and Technology, 2008).

These integrated and reliable renewable energy systems are not limited to Germany or Europe. In Zambia, an interconnected solar–biomass–micro–hydro network will generate base-load electricity for a collection of local villages. The combined system will include one biomass power plant, one micro-hydroelectric station, and a collection of distributed solar panels with a collective output of 2.4MW, and it is expected to begin operation in 2010 (United Nations Industrial Development Organization, 2009). In Cuba, a hybrid biomass gasification power plant, four distributed biogas plants, and one wind farm will have a rated capacity of 11MW and begin generating base-load electricity for the Isla de la Juventud in

2011 (United Nations Industrial Development Organization, 2009). In the village of Xcalak, Mexico, 234 solar panels have been integrated with 36 batteries, 6 wind turbines, a 40kW inverter to convert DC power to AC, and a sophisticated control system. The system has so far displaced the need to construct a US$3.2 million transmission line extension, and in its first year of operation proved more reliable than the diesel generators that it replaced, although one is still installed as a backup just in case (US Department of Energy, 2006).

Other hybrid systems operate at the scale of the individual home or business. In Canon City, Colorado, a joint wind–solar system meets almost the entire home's power needs. The system includes 24 120W solar PV modules, 20 batteries, 2 small-scale wind turbines, 2 tilt-up wind turbine towers and a vented battery box (US Department of Energy, 2006).

At the commercial headquarters of the Rocky Mountain Institute, solar thermal and solar panels have been integrated with energy efficiency techniques so that renewable energy meets 99 per cent of the building's demand for energy services. Using passive design, super efficient windows, state-of-the-art insulation and ventilation, a solar water heater, two exterior tracking solar photovoltaic systems, and a collection of stationary flat-plate solar panels integrated into the roof, the building needs no conventional heating systems (see Figure 7.2). The building is so efficient, when one of the owners wants to turn up the heat, he or she jokingly remarks that it can be harnessed 'from a 50W dog, adjustable to 100W by throwing a ball' (Sovacool, 2008).

Similarly, a Californian prison has integrated about three acres of solar panels with net metering, energy efficiency, and storage systems to not only meet the facility's energy needs but also export electricity back to the grid. The entire system

Figure 7.2 *The ultra-efficient Rocky Mountain Institute headquarters in Snowmass, Colorado*

Source: Sovacool, 2008

is designed to collect, store and save electricity in the morning and at night, and to then sell it back to the grid during times of peak demand, when it is worth the most. The system paid for itself within the first year and has since brought in hundreds of thousands of dollars of additional savings (Sovacool, 2008).

Clearly, such hybrid renewable systems appear to be possible in virtually any location and at almost any scale.

Looking to the future, as the performance of batteries improves and their costs drop, they too could begin to back up large amounts of solar and wind power. Similarly, the greater use of plug-in hybrid electric vehicles and vehicle-to-grid technologies can also enhance the competitiveness of intermittent renewables, as these innovations would enable automobiles to store energy from wind and solar that can then be recalled when needed (Michel, 2007). The potential resource base for tapping batteries in vehicles to store energy from renewables is staggering. Placing just a 15kW battery in each of the existing 191 million automobiles in the US would create 2865GW of equivalent electricity capacity if all the vehicles supplied power simultaneously to the grid, a capacity more than twice that of the entire existing electricity industry, which has slightly more than 1000GW of installed capacity (Sovacool and Hirsh, 2009).

7.5 Storage Technologies for Backup

When interconnecting wind and solar farms or creating hybrid systems is not practicable or possible, intermittent renewable systems can be integrated with energy storage technologies to eliminate their intermittency. While batteries are the most common type of commercially available storage technology on the market, they are also the most expensive and tend to store only small amounts of energy. Three other types of storage systems – pumped hydro, compressed air, and molten salt – offer a cheaper alternative and can handle larger amounts of capacity. Each of these three options tends to add only an extra US$0.006–0.051/kWh to the levelized cost of renewable electricity generation.

One of the most cost-effective, and widely used, storage technologies to smooth out the intermittency of wind and solar is pumped hydro systems. These storage systems generate electricity by reversing flow between two water sources, often elevated reservoirs or water towers. They can use the electricity from wind and solar during the day (or over any period of time) to pump and store water, and then release the water during the night (or any period of time) to turn a generator that produces electricity (see Figure 7.3). Bonneville Power Administration, a large federal utility in the Pacific Northwest, uses its existing 7000MW hydroelectric and pumped hydro storage network to do this. Starting in 2005, Bonneville offered a new business service to 'soak up' any amount of intermittent renewable output, and sell it as firm output from its hydropower network one week later. Such storage technologies can have greater than 1000MW of capacity (depending on location),

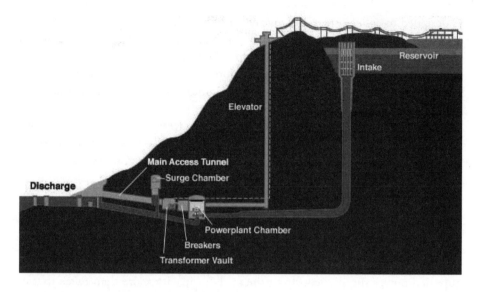

Figure 7.3 *A typical pumped hydro storage system*

Source: US Tennessee Valley Authority, 2005

and operate according to fast response times and relatively low operating costs. The Helms Pump Storage Facility near Fresno, California, for example, has three units totalling 1200MW of generation capacity. Worldwide more than 90GW of pumped hydro storage facilities operated in 2007 (California Independent System Operator, 2008).

Compressed air energy storage (CAES) is also economical for large bulk storage. Renewable CAES systems use intermittently generated renewable energy to compress air and pump it into underground formations such as caverns, abandoned mines, aquifers and depleted natural gas wells. The pressurized air is then released on demand to turn a turbine that generates electricity (see Figure 7.4). Land and expertise are widely available for renewable CAES systems. Sufficient CAES resources are available in 75 per cent of the US and operators already have an 80-year history of storing pressurized natural gas in underground reservoirs (Fthenakis et al, 2009). Existing CAES plants still use small amounts of natural gas to heat compressed air, but consumption is 60 per cent lower than single-cycle natural gas turbines. Natural gas fuel needs are set to be eliminated entirely in the newest designs relying on advanced adiabatic CAES, which are expected to be commercialized as early as 2015. Paul Denholm and his colleagues from the National Renewable Energy Laboratory in the US, for example, note that attaching wind turbines to CAES technologies can improve their capacity factor above 70 per cent, making them 'functionally equivalent to a conventional baseload plant' (Denholm et al, 2005).

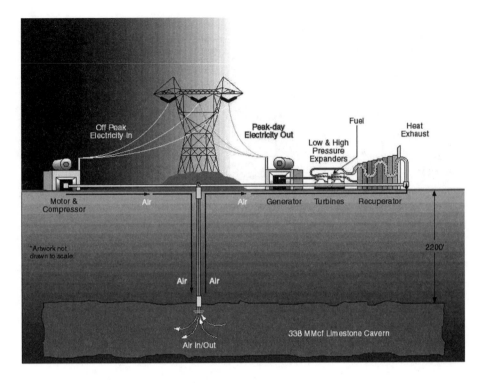

Figure 7.4 *A typical compressed air energy storage system*

Source: Sandia National Laboratory, 2001

Molten salt storage facilities also work well with renewable energy technologies that generate a lot of heat, such as geothermal and bioelectric plants or concentrated and solar thermal systems. Molten salt storage facilities store the extra heat generated from hydrothermal fluids (for geothermal units), combustion of crops and waste (biomass), or concentrated sunlight (solar thermal) in large insulated tanks filled with molten salt (see Figure 7.5). These tanks are very efficient in retaining heat, which can then be stored and used to produce electricity later. Typical tanks store enough energy to power a 100MW turbine for 2–12 hours; when many tanks are combined they can store enough energy to power that turbine for more than a week, with an efficiency loss of less than 1 per cent (Sena-Henderson, 2006). Arizona Public Service Company, for instance, uses molten salt storage at their 280MW Solana Generating Station near Phoenix in Gila Bend, Arizona. There, three square miles of parabolic troughs transfer heat to molten salt storage tanks that then power two 140MW steam generators, creating an entirely base-load solar plant (Lockwood, 2008). A similar solar–salt base-load system in Andasol, Southern Spain, has been operating reliably since 2008.

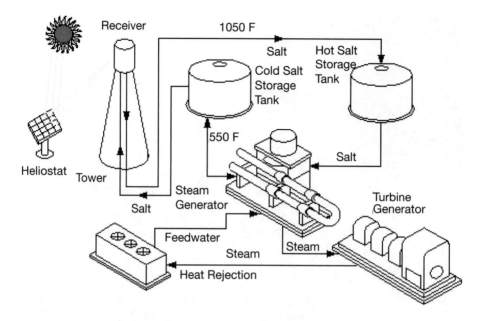

Figure 7.5 *A typical molten salt storage facility*

Source: Sena-Henderson, 2006

7.6 Marginal Grid Integration and Transmission Costs

Opponents to renewable energy and FITs sometimes state that since distributed renewable resources are so diffuse and remote from users, integrating them into the grid and constructing transmission and distribution lines to them will be prohibitively expensive. Apart from ignoring that many small-scale renewable units will be decentralized, integrated into buildings, and close to end-users, such a claim also ignores a preponderance of recent evidence from the US and around the world.

Consider the actual costs of grid integration. One independent system operator in the US projected the extra costs of operating 152 separate wind farms in the Midwest, each one 40MW, and calculated their operation every five minutes as the simulation progressed over three years. The study found that the additional cost of reserves for managing the overall system were about $0.001/kWh, and that the wind farms also improved system reliability (EnerNex Corporation and Midwest Independent System Operator, 2006). Researchers from Stanford University believe the cost to be even lower. They estimated that the utility expense of providing ancillary service to adjust to the intermittency of dispersed wind turbines was about $0.00005–0.0003/kWh, less than 1 per cent of the price of producing wind electricity (Jacobson and Masters, 2001).

Another study examined five separate electric utilities with wind penetration rates from 10–29 per cent of their total capacity, and found an extra operating cost impact of only about $0.004/kWh. The study concluded that wind penetrations up to 20 per cent of system peak demand increased system operating costs by no more than $0.005/kWh and never exceeded 10 per cent of the system operating costs (Smith et al, 2007, pp900–908). Interestingly, the study also found that the total operating cost for these wind farms *including the need for backup and transmission* was $0.019–0.0497/kWh, cheaper than virtually any other source of electricity on the market (Smith et al, 2007, p904). Yet another international comparison of 15 studies examining balancing and operating costs with wind penetration of 10–25 per cent of gross demand in Finland, Denmark, Germany, Ireland, Norway, Portugal, Spain, Sweden, the UK, and the US again found integration costs ranging from a mere US$0.003–0.005/kWh (Holttinen et al, 2007). Clearly, the extra costs for integrating dispersed and remote renewable resources are only a small fraction of their overall cost, and have not hurt the economic competitiveness of actual projects already operating in many parts of the US and Europe.

Now consider the cost of transmission. A recent survey of transmission studies in the US looked at 40 different estimates from 2001 to 2008 of building transmission and distribution lines to remote wind farms across a broad geographic area of the country. The study noted that the median cost of transmission, including building lines to some of the most remote locations, would be only about $0.015/kWh (Mills et al, 2009). In all likelihood, the true costs could be much lower. New lines need not be built from scratch, and can piggyback on existing transmission towers, and smaller lines on existing towers can be upgraded to minimize concerns over rights of way and siting. Only local connections to the nearest long-distance pathways will need new lines, and these will be rare indeed. Remaining with the US as an example, since its land area is much larger than individual European countries, a huge 175,000MW of wind power potential exists within 8 kilometres of existing 230kV or lower transmission lines and 840,000MW lies within 20 kilometres of existing transmission lines (Jacobson and Masters, 2001). If remote wind locations were fully developed in the US, the cost of above-ground transmission lines would run to about $310,000 per kilometre, meaning the cost of 10,000 kilometres of new lines to these remote locations could be $3.1 billion, less than 1 per cent of the cost of building the 225,000 new turbines they would connect to (Jacobson and Masters, 2001). Put another way, at a distance of 2000 kilometres, the cost of building new high voltage transmission lines to remote wind farms is estimated to add a mere $0.007/kWh to the levelized cost of a new wind project (Jacobson and Masters, 2001).

7.7 Conclusion

In short, the supposedly technical reasons against the rapid expansion of renewables, commercial or small-scale, does not hold up to reason or operating experience. Conventional fossil-fuelled and nuclear facilities suffer a host of reliability problems relating to unplanned outages, long construction lead times, unexpected changes in demand for electricity, and long-distance transmission from centralized locations to decentralized users. The renewable resources promoted by FITs respond effectively to each of these challenges: they have a higher technical reliability, shorter construction lead times, can be quickly installed in modular increments to meet any scale of electricity demand, and small-scale systems can be utilized close to end-users.

Furthermore, not all of the renewable resources promoted by FITs are variable and intermittent – biomass, biogas, geothermal, solar thermal, and hydroelectric facilities operate just as predictably as conventional units. Variable renewable resources become collectively reliable when diversified by fuel and location, and even more reliable when integrated with other renewable technologies or coupled with batteries, pumped hydro facilities, compressed air energy storage or molten salt storage. When necessary, the costs of grid integration and the construction of new transmission and distribution lines represent a very small fraction of total project costs. In the US, grid integration usually adds less than $0.01/kWh to the levelized costs of wind electricity and transmission, less than 1 per cent of the total costs for a remote wind project. As the International Energy Agency, hardly an organization biased in favour of renewables, recently argued, 'variability will rarely be a bar to increased renewables deployment' (Olz et al, 2007, p5). Perhaps such a statement needs to be modified to say that variability will only prevent increased renewables deployment so long as people mistakenly think it will.

References

Archer, C. L. and Jacobson, M. Z. (2007) 'Supplying baseload power and reducing transmission requirements by interconnecting wind farms', *Journal of Applied Meteorology and Climatology*, vol 46, pp1701–1717

AWS TrueWind (2005) 'An analysis of the impacts of large-scale wind generation on the Ontario electricity system', Canadian Independent Electricity System Operator, 26 April, www.uwig.org/IESO_Study_final_document1.pdf

Bertani, R. (2002) 'What is geothermal potential?', *IGA News*, vol 53, pp1–3

California Independent System Operator (2008) 'Integration of energy storage technology in power systems', Northwest Wind Integration Forum Pumped Hydro Storage Workshop, p2, www.nwcouncil.org/energy/wind/meetings/2008/10/DavidHawkins.pdf

Cavanagh, R. (1986) 'Least-cost planning imperatives for electric utilities and their regulators', *Harvard Environmental Law Review*, vol 10, pp299–344

DeMeo, E. A., Grant, W., Milligan, M. R. and Schuerger, M. J. (2005) 'Wind plant integration: Cost, status, and issues', *IEEE Power and Energy Magazine*, vol 3, no 6, pp39–46

Denholm, P. (2006) 'Improving the technical, environmental and social performance of wind energy systems using biomass-based energy storage', *Renewable Energy*, vol 31, p1356

Denholm, P., Kulcinski, G. L. and Holloway, T. (2005) 'Emissions and energy efficiency assessment of baseload wind energy systems', *Environmental Science and Technology*, vol 39, pp1903–1911

EnerNex Corporation and the Midwest Independent System Operator (2006) *Final Report: 2006 Minnesota Wind Integration Study*, EnerNex Corporation, Knoxville, TN

European Wind Energy Association (2005) *Large Scale Integration of Wind Energy in the European Power Supply: Analysis, Issues and Recommendations*, EWEA, Paris, p13

Federal Ministry of Economics and Technology (2008) *E-Energy: ICT-Based Energy Systems of the Future*, BWMi, Berlin

Fthenakis, V., Mason, J. E. and Zweibel, K. (2009) 'The technical, geographical, and economic feasibility for solar energy to supply the energy needs of the US', *Energy Policy*, vol 37, pp387–399

Gul, T. and Stenzel, T. (2006) 'Intermittency of wind: The wider picture', *International Journal of Global Energy Issues*, vol 25, p173

Harvey, W. (2008) quoted in 'Renewable energy: Price and policy are key', *Environmental Research Web*, 30 July

Holttinen, H., Lemström, B., Meibom, P., Bindner, H., Orths, A., Van Hulle, F., Ensslin, C., Tiedemann, A., Hofmann, L., Winter, W., Tuohy, A., O'Malley, M., Smith, P., Pierik, J., Tande, J. O., Estanqueiro, A., Gomez, E, Söder, L., Strbac, G., Shakoor, A., Smith, J. C., Parsons, P., Milligan, M. and Wan, Y. (2007) 'Design and operation of power systems with large amounts of wind power, State-of-the-art report', VTT Working Papers 82, VTT, Espoo, Finland. Presented at the European Wind Energy Conference and Exhibition (7–10 May), Milan, Italy, www.vtt.fi/inf/pdf/workingpapers/2007/W82.pdf

International Hydropower Association, International Commission on Large Dams, Implementing Agreement on Hydropower Technologies and Programmes/International Energy Agency, Canadian Hydropower Association (2000) 'Hydropower and the World's Energy Future', www.ieahydro.org/reports/Hydrofut.pdf

International Energy Agency (2005) *Variability of Wind Power and Other Renewables: Management Options and Strategies*, International Energy Agency, Paris, p20

Jacobson, M. Z. and Masters, G. M. (2001) 'Letters and responses: The real cost of wind energy', *Science*, vol 294, no 5544, pp1000–1003

Lockwood, B. D. (2008) *Blowing in the Wind: Renewable Energy as the Answer to an Economy Adrift*, Testimony before the House Select Committee on Energy Independence and Global Warming, 6 March, US Government Printing Office, Washington, DC

Lovins, A. B., Sheikh, I. and Markevich, A. (2008) 'Forget nuclear', *Rocky Mountain Institute Solutions*, vol 24, no 1, pp23–27

Michel, J. H. (2007) 'The case for renewable FITs', *Journal of EUEC*, vol 1, pp2–19

Mills, A., Wiser, R. and Porter, K. (2009) *The Cost of Transmission for Wind Energy: A Review of Transmission Planning Studies*, LBNL-1471E, Lawrence Berkeley National Laboratory, Berkeley, CA

NERC (2005) *2000–2004 Generating Availability Report*, www.nerc.com/_gads/

Olz, S., Sims, R. and Kirchner, N. (2007) *Contributions of Renewables to Energy Security: International Energy Agency Information Paper*, OECD, Paris

Perin, C. (1998) 'Operating as experimenting: Synthesizing engineering and scientific values in nuclear power production', *Science, Technology, and Human Values*, vol 23, no 1, pp98–128

Piwko, R., Osborn, D., Gramlich, R., Jordan, G., Hawkins, D. and Porter, K. (2005) 'Wind energy delivery issues: Transmission planning and competitive electricity market operation', *IEEE Power and Energy Magazine*, vol 3, no 6, pp47–56

Sandia National Laboratory (2001) 'Sandia assists with mine assessment', www.sandia.gov/media/NewsRel/NR2001/images/jpg/minebw.jpg

Sena-Henderson, L. (2006) *Advantages of Using Molten Salt*, Sandia National Laboratory, Albuquerque, NM

Smith, J., Milligan, M., DeMeo, E. A. and Parsons, B. (2007) 'Utility wind integration and operating impact state of the art', *IEEE Transactions on Power Systems*, vol 22, no 3, pp900–908

Sovacool, B. K. (2008) *The Dirty Energy Dilemma: What's Blocking Clean Power in the United States*, Praeger, Westport, CO, p85

Sovacool, B. K. (2009) 'The intermittency of wind, solar, and renewable electricity generators: Technical barrier or rhetorical excuse?', *Utilities Policy*, vol 17, no 3, September pp288–296

Sovacool, B. K. and Hirsh, R. F. (2009) 'Beyond batteries: An examination of the benefits and barriers to plug-in hybrid electric vehicles (PHEVs) and a vehicle-to-grid (V2G) transition', *Energy Policy*, vol 37, no 3, p1096

Topfer, K. (2006) *Background Paper: The Combined Power Plant*, Erneurbare Energien, www.kombikraftwerk.de

Trade Wind (2009) *Integrating Wind: Developing Europe's Power Market for the Large-Scale Integration of Wind Power*, European Renewable Energy Council, Brussels

US Department of Energy (2006) *PV in Hybrid Power Systems*, Office of Energy Efficiency and Renewable Energy, Washington, DC, p1

United Nations Industrial Development Organization (2009) *UNIDO and Renewable Energy: Greening the Industrial Agenda*, UNIDO, Vienna, pp20–21

US Congressional Budget Office (2008) *Nuclear Power's Role in Generating Electricity*, DBO, Washington, DC, p17

US Tennessee Valley Authority (2005) 'Pumped storage plant', www.tva.gov/power/images/pumpstor.jpg, accessed April 2009

Vine, E. D., Kushler, M. and York, D. (2007) 'Energy myth ten: Energy efficiency measures are unreliable, unpredictable, and unenforceable', in: B. K. Sovacool and M. A. Brown (eds) *Energy and American Society – Thirteen Myths*, Springer, New York, NY

Wenger, H. J., Hoff, T. E. and Farmer, B. K. (1994) 'Measuring the value of distributed photovoltaic generation: Final results of the Kerman grid-support project' presentation at the First World Conference on Photovoltaic Energy Conversion Conference, Waikaloa, Hawaii, December 1994, IEEE, Washington, DC, pp792–796

Winfield, M. S., Cretney, A., Czajkowski, P. and Wong, R. (2006) 'Nuclear power in Canada: An examination of risks, impacts, and sustainability', Pembina Institute, http://ontario.pembina.org/pub/1346, p4

8

Barriers to Renewable Energy Deployment

The question often arises – if renewable electricity technologies have such impressive benefits, why do we need feed-in tariffs (FITs) to promote them? Why aren't renewables automatically endorsed by the marketplace, adopted by utilities, accepted by businesses and purchased by homeowners?

The answer is that the evolution of technology, especially energy and electricity systems, does not occur in a Darwinian world where all possibilities begin on an equal footing, and then are accepted and rejected based solely on merit or efficiency. The ancient philosopher Heraclitus, writing at the time of Plato, once said that 'you can never step twice into the same river'. His statement implies that the river is forever altered from your first step into it. In the same way, more than one hundred years of decisions and government subsidies for conventional energy technologies have distorted the 'free market' for electricity much like Heraclitus' foot forever distorted the river.

Every single energy system in use today has required government intervention to overcome a web of obstacles, barriers, impediments and challenges. The diffusion of energy technologies, from nuclear reactors to compact fluorescent light bulbs, has historically been a slow process, often taking decades and in some cases centuries. In an analysis of the consumer acceptance of 20 different energy technologies, Peter Lund found that market penetration for novel innovations can take anywhere from fewer than 10 years to more than 70 years (Lund, 2006, 2007). The diesel engine, for example, needed about 60 years before it became fully embraced as a commercial technology. An assessment of the barriers to all greenhouse gas-reducing technologies, not just renewable power plants but also energy efficiency, biofuels and others, found that incumbent technology support systems can 'lock in' conventional technologies even in the face of superior substitutes. The study also noted that transaction costs, such as gathering and processing information, developing patents, obtaining permits and designing contracts, can all be prohibitive during the early stages of a technology's development (Brown et al, 2008). These impediments also exist in the developing world, where another global survey found that stakeholders lack the proper information about alternative energy systems, and without policy intervention these will not achieve wider use (Reddy and Painuly, 2004).

When it comes to the barriers to renewable electricity and FITs, obstacles fall generally into four categories, each explored in this chapter:

1 *Financial and market* impediments include the lack of readily available information on FITs to both users and producers; improper discount rates and unacceptably high rates of return for other energy investments; the principal–agent problem; predatory practices undertaken by some energy firms and electric utilities; and a desire for businesses and industries to stick to their core missions rather than invest in new forms of energy supply.
2 *Political and regulatory obstacles* encompass inconsistent government standards and fragmented policy making and a tendency to subsidize fossil fuels and nuclear power technologies over renewables.
3 *Cultural and behavioural barriers* relate to public misunderstanding about electricity; public expectations about cheap and abundant forms of electricity supply; and a strong personal desire among consumers to prioritize comfort, control and freedom, rather than sustainability.
4 *Aesthetic and environmental challenges* include aesthetic values and the sometimes symbolic nature of FITs and renewable energy projects.[1]

8.1 Financial and Market Impediments

Perhaps one of the simplest barriers to both FITs and renewable energy systems is lack of information. A bunch of smart economists, many of them winning Nobel Prizes, have been arguing for the past half century that the production and consumption of information will not occur sufficiently in most markets without intervention and education (Stigler, 1961; Akerlof, 1970). The provision of information is subject to what these economists have called a public goods problem because the production of useful information is valuable to everyone, not just to the person who produced it. Furthermore, those that have information may have strategic reasons to manipulate its value. Sellers may intentionally give misinformation to make their products seem more attractive; and the costs of acquiring reliable information may be significant, especially when up against well-distributed misinformation for example by oil companies or anti-environmental groups.

A supplier of faulty solar panels, dodgy wind turbines, or fossil fuels will often possess 'better' information than potential buyers and users, so they can deceive customers, leading to a reluctance of consumers to trust even an honest or alternative seller's claims. For instance, one group looking at the barriers to small-scale renewables (and other distributed systems) in the Southern part of the US found that an information asymmetry existed: where stakeholders were familiar with unconventional power sources such as wind and solar, the negative experiences were the best known (Southern States Energy Board, 2003). In Germany, public resistance to wind energy can be found in regions with no or very limited wind power capacity – apparently due to misinformation. By contrast, people in regions with a large number of wind turbines are generally less critical.

Producers face information failure about renewables and FITs as much as consumers. One study of American utility executives found that they had very little knowledge of customer demand, tastes or preferences. After the Harvard Business School convened a conference on New England's power needs, for instance, sponsors publicly denounced the:

> *shared ignorance among some of the most informed people in the country about some very fundamental things concerning the nature of electrical energy demand. The managers appeared to know less about electricity demand than other business executives knew about the nature of demand for toothpaste, lifesavers, or beer.* (Nakarado, 1996)

A related impediment to lack of information is lack of capital. Many homeowners simply do not have the resources to purchase their own small-scale wind turbine or solar panel, or to invest in a community energy project. Once people with fixed incomes spend money on anything (the Beatles' *White Album*, stock options, a new television, an electronic can opener), they have already 'sunk' their available earnings. A recent survey found that those that did install solar panels on their homes did so mostly through cash, yet 60 per cent of consumers interviewed said they do not have any cash available for such investments (Bulat et al, 2008). Moreover, those who often could benefit from renewable energy the most (i.e. the poorest) have the least money to invest in it. This barrier can be overcome by FITs. The secured income from tariff payments allows people even without savings or cash to get a loan from the bank in order to buy a solar panel or wind turbine.

Closely connected to lack of capital is a concept known as the discount rate, or how consumers make investment decisions when they do have capital available. An implicit discount rate refers to the rate at which consumers want to recover their investment in a given item. One survey asked consumers about the payback times expected for investments in energy systems, and found that one-third of the respondents were unable to answer the question at all. Of those that answered, more than three-quarters indicated that they would not invest unless they received payback on their investment in three years or less (Koomey, 1990).

One study found a general aversion against solar PV systems in the housing market for precisely these reasons (Barbose et al, 2006). Such systems greatly add to the initial cost of purchasing a home. Moreover, when times are good and houses are selling well, builders and real estate agents view it as an indicator that alternative energy technologies are not needed to make sales. When times are bad, they place even more emphasis on minimizing costs and keeping house prices low. Homeowners also worry about project delays (and thus rising costs) associated with PV availability, installation scheduling, and utility interconnection. Builders believe that most homebuyers are not interested in PV, given its extra cost, and that many may even be opposed to it due to concerns about aesthetics, maintenance, or reliability.

A 2008 study from the Harvard Business School of hundreds of builders and contractors is most telling here. When asked how they would spend an extra $10,000 in the construction budget on discretionary items, more than 25 per cent interviewed said they would do granite counter tops while less than 15 per cent said solar panels. The reasons had to do with the fact that granite counter tops were perceived as less risky and more visible than solar panels (Bulat et al, 2008). The survey also found that consumers had unrealistic expectations about payback, many expecting a $4000 investment in solar PV to save more than 50 per cent on monthly utility bills, when in reality it would tend to save 10–15 per cent (Bulat et al, 2008).

Another economic barrier is known as the principal–agent problem, or when those making investment decisions (the principals) do not have to live with the results (the agents). The principal–agent problem comes in many different flavours relating to renewable energy investments. Architects, engineers and builders design homes that they will not live in. Landlords purchase equipment for tenants that they will not use themselves. Industrial procurers select technology for their plants. A severe misalignment occurs when consumers use technologies selected by others, especially when intermediaries overemphasize up-front cost rather than life-cycle costs. Four sets of split incentives relating to renewable energy and FITs appear to be the most pernicious: distinctions between builders and homeowners, between landlords and tenants, within business investors, and within utilities.

Architects, engineers, and builders select energy technologies that homeowners and dwellers use. Yet the prevailing fee structures for building design are based on a percentage of capital cost of a project, penalizing engineers and architects for installing efficient but more expensive renewable energy systems. The pressure to lower first costs is reinforced by banks, lenders and financiers, since the builder is, in effect, building the house for them, and their criteria for selling a loan include keeping the ratio of monthly payments to monthly income low enough to make the loan a reasonable risk (Anderson, 1995; Brown, 1993, 1997).

Developers and investors typically want fast, cheap buildings up and running quickly so they can start maximizing returns. They do not want to purchase solar panels and wind turbines which would raise project costs. Thankfully, FITs have the potential to turn this impediment on its head since residential-scale renewable resources would become assets to make the homeowner money, rather than liabilities.

As a related barrier, tenants have no interest in investing in renewable energy, since they do not own the property and may have short-term occupancy; landlords don't because they can pass energy costs on to tenants, and retrofits often appear risky and unprofitable. This classic problem of 'split incentives' explains why energy consumption and expenditure per unit of floor area are much greater in rented buildings and public housing than in owner-occupied single-family housing (DeCicco et al, 1996).

The same types of split incentives occur in some of the country's most energy-intensive industries. Many industrial facilities have only one utility meter to measure plant-wide consumption. In these situations, traditional accounting practices treat energy as an overhead cost, which then becomes allocated across departments according to their numbers of workers or square feet of floor space. The drawback is that the cost of any one department's energy waste or inefficiency is distributed to all departments. Conversely, any department that undertakes investments in renewable energy will have its improvements diluted by the artificial allocation of costs (Howarth et al, 2000).

In a similar vein, most small and large businesses resist investing in and using renewable energy because these technologies are believed to deviate from each company's core business mission. For the typical business, energy costs are a miniscule fraction of labour costs; therefore, management and capital are drawn to other areas. Even though these businesses obviously use electricity, they seem to have little to no interest in producing power. Because most non-utility and non-energy businesses have goals and priorities that have nothing to do with electricity, they tend to be concerned with the promotion of their own corporate strategy. The interests and technologies such businesses promote are not interoperable with investments in renewable energy, and so they refuse to express interest and commitment.

As a final economic barrier, non-commercial-scale renewable resources (such as small-scale wind, solar PV, geothermal heat pumps, and solar thermal water heaters) directly threaten the market share of electric utilities, energy companies and other power operators. These incumbent providers often have a monopoly on all or part of the electricity system, owning generators, transmission lines or distribution stations, or in some cases all three. Ordinary consumers have little ability to counteract the market power held by dominant electricity providers and distributors. The best obstacle to price discrimination is to buy from other sources, but most consumers have only a few suppliers. Resale by consumers is implausible, distribution is controlled by central stations and vertically integrated with production, and electricity cannot be stored effectively. Customers have no practical way of reselling the product, meaning they have little capacity to hedge against the monopoly power of dominant utilities (Yakubovich et al, 2005).

Analogously, many large energy companies have actively used their 'power of incumbency' to further mould government regulations in favour of large, centralized plants (and disadvantage small, decentralized units). A few independent studies confirmed that system operators have attempted to retain their control over the electric utility system by employing a wide variety of predatory and discriminatory practices. Such efforts typically begin with the imposition of fees to connect to the grid (see Section 2.10). In some countries and states that have begun restructuring or 'deregulating' their utility systems, formerly regulated 'natural monopoly' power companies have been permitted to charge customers 'stranded costs' (Allen, 2002). These costs are intended to cover a 'fair return' on generation and transmission investments made by utilities during the era of regulation, when the investments were viewed as serving all users. Put simply, when a customer decides to install a solar panel or wind turbine independent from the utility, he or she arguably removes part of the grid's existing load requirement and 'strands' part of the investment the utility made in the power system. Stranded costs allow the utility to continue to charge the consumer for these investments, even if they don't use the system.

Other discriminatory practices include charging demand fees (a charge that penalizes customers for displacing demand from utilities). A recent study undertaken in the US found more than 17 different 'extraneous' charges associated with the use of dispersed renewable technologies (Alderfer and Starrs, 2000). These types of charges, the senior editor of *Public Utilities Fortnightly* exclaimed, 'are a major obstacle to the development of a competitive electricity market' (Stavros, 1999).

8.2 POLITICAL AND REGULATORY OBSTACLES

Political and regulatory obstacles play their part as well. The most obvious barrier relates to the inconsistent political support for renewable energy systems. Unlike subsidies and incentives for fossil-fuelled technologies, policies aimed at encouraging renewables have changed frequently, greatly discouraging widespread adoption of the technologies.

The transition from the Jimmy Carter Administration to the Ronald Reagan Administration in the US, for example, financially endangered government research for distributed and renewable energy resources, and drove some people out of the renewable and small-scale energy industry altogether. Wilson Prichett, a consultant who has managed more than 300 renewable energy projects in the US, argued that President Reagan discouraged everyone, not just policy makers in the government, from working on energy systems. As a result of Reagan's policies, Prichett recalled that:

> By 1982 or 1983, most everyone had left the industry. All those thousands of people around the country working on ways to convert agricultural waste to fuels and different solar and wind designs just stopped doing it. There was no political encouragement. In fact, there was complete discouragement.[2]

As a glaring sign of such discouragement, consider the sad tale of production tax credits for biomass, wind and solar in the US. The Energy Policy Act of 1992, signed by Reagan's successor, George H. W. Bush, provided a production tax credit for certain renewable energy technologies. Those credits expired in 1999, and environmental advocates worked diligently to win Congressional approval for their reinstatement, often on an annual basis. When Congress failed to restore the credits before the end of 2001, investment in wind turbine projects dropped precipitously. Developers installed only 410MW of new wind turbines in 2002, down from about 1600MW in 2001 and 2003. The conflicting policies for wind turbines in the US have created boom and bust cycles within the industry, making it all but impossible to obtain financing for projects.

Policy change had an equally disruptive effect on the Danish wind power market in 2001. At that time, the government decided to replace their successful FIT scheme with a tradable green certificate scheme. Even though this alternative was never fully implemented, new wind installations of power capacity in Denmark fell from 600MW in 2000 to merely 18MW in the first half of 2001 as investors grappled with the uncertainty surrounding possible regulatory changes (see Figure 8.1) (Mendonça et al, 2009).

As any renewable energy investor, manufacturer or operator already knows, the variability of policy relating to renewable energy technologies serves as a serious impediment. Entrepreneurs seeking investment from individuals and institutions often require consistent conditions upon which to make decisions. Forecasts of profitability usually require data concerning tax credits, depreciation schedules, cash flows and the like, well into the future. When policy makers frequently change

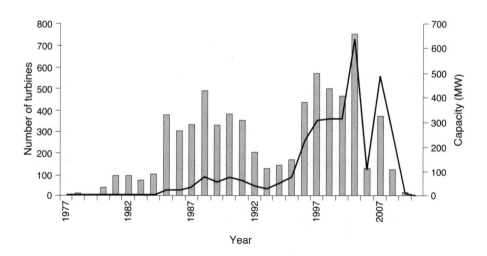

Figure 8.1 *Annual growth of installed wind power capacity in Denmark*

Source: Meyer, 2007

the factors that go into these financial calculations, they insert an extra level of uncertainty into the decision-making process.

Such impediments are not limited to the US. In the EU, a recent study surveying 21 countries found a host of political barriers which created considerable uncertainty (Coenraads et al, 2008). Red tape and unforeseeable bureaucratic delays from local and national authorities were cited by project developers as a significant impediment, with the average renewable energy project involving more than nine separate authorities. Long lead times for project authorization were also mentioned as a problem, along with lack of political support to create harmonized standards and policies so small-scale renewables could be interconnected to the grid.

For example, the licensing process for renewable energy projects is often prolonged by the acquisition of the necessary planning permits. Due to the decentralized nature of renewable energy projects and the frequent necessity of land use, anachronistic planning provisions can slow down the permitting process. Small-scale hydropower projects in Austria for instance, have to go through a permitting process which can take up to 12 months, but in Spain and Italy the permitting process sometimes takes up to 12 years. In the case of wind energy, the administrative process in Germany only takes one or two years, while the complicated French procedure sometimes takes up to five years (Coenraads et al, 2006).

Another obstacle for renewable energy projects is the large number of authorities that have to be contacted for a large variety of permits, including the industrial plant procedure, the grid connection procedure and the environmental assessment. For the installation of a wind power plant in France, the producer will have to contact 27 different authorities at different political levels. This does not only unnecessarily lengthen renewables projects, it also increases the overall project costs and consequently the additional burden on the final electricity consumer. In Hungary, on average 40 authorities are involved in the permitting process (Coenraads et al, 2008).

Examples also abound in the US. One in-depth study of the 'red tape' involved in renewable energy projects found that systems installers frequently faced planners and buildings inspectors with little to no experience of permitting solar and wind systems (Pitt, 2008). The study noted that complex permitting requirements and lengthy review processes, much like some of the European countries, result in project delays and substantially add to the costs of projects. Multiple permitting standards across jurisdictions, such as competing or convoluted city, county, state and national building codes, only add to the complexity. The study concluded that 'these remaining bureaucratic hurdles stymie efforts by homeowners and business owners to install systems and hinder development of a national market for distributed renewable energy systems' (Pitt, 2008).

Other examples of administrative barriers are presented in Section 2.11.

Yet the most egregious, albeit often invisible, impediment concerns continued government subsidies for conventional resources, often directly at the expense

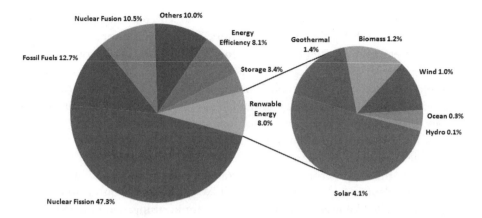

Figure 8.2 *Energy research subsidies in OECD countries, 1974–2002*

Source: International Energy Agency, 2004

of subsidies for renewables. Despite all of the talk about climate change and promoting clean energy, existing subsidies continue to heavily favour nuclear power and fossil fuels. Funding for renewable energy as a whole peaked in 1980 at $2.1 billion, then dropped to $750 million in countries of the Organisation for Economic Co-operation and Development (OECD). Nuclear energy on the other hand received close to 70 per cent of all energy-related research expenditures from 1974 to 2006 followed by coal, oil and gas (International Energy Agency, 2008). Put another way, from 1974 to 2002 nuclear power received $138 billion in subsidies from OECD countries and fossil fuels an additional $37 billion, but solar photovoltaics received $6.3 billion and wind energy a miserly $2.9 billion (See Figure 8.2).

In many industrialized countries, especially the US, coal producers still receive a percentage depletion allowance for mining operations, deductions for mining exploration and development costs, special capital gains treatment for coal and iron ore, a special deduction for mine reclamation and closing, research subsidies, and black-lung benefits paid for by national governments (black-lung disease, or coal worker's pneumoconiosis, is caused by long-term exposure to coal dust and causes fibrosis, necrosis and lung failure). Oil and gas producers still receive a similar depletion allowance, bonuses for enhanced oil recovery, tax reductions for drilling and development costs, fuel production credits, and research subsidies (Jacobson and Masters, 2001). Nuclear energy operators and manufacturers benefit from massive loan guarantees, research funds, public insurance and compensation against construction delays, tax breaks for decommissioning, tax credits for operation, and government-funded off-site security and nuclear waste storage.

Existing subsidies for nuclear power in the US, for instance, are nothing short of obscene. Historically, nuclear power development received subsidies worth

$15.30/kWh its first 15 years of development, which compares with subsidies worth only $7.19/kWh for solar and $0.46/kWh for wind during their first 15 years of development. Taken as a whole, from 1947 to 2000 cumulative subsidies for nuclear power amounted to $1411 per US household, compared to just $11 per household for wind (Sovacool, 2009a, pp1529–1541). The two most recent legislative acts concerning energy in the US are filled with subsidies for nuclear energy, including $13 billion worth of loan guarantees covering up to 80 per cent of project costs, $3 billion in research funds, $2 billion of public insurance against delays and $1.3 billion in tax breaks for decommissioning. As if this was not enough, the US government also provides an extra $0.018/kWh in operating subsidies for the first eight years and up to 6GW constructed (equivalent to about $842 per installed kilowatt), government funds for licensing, and limited liability for accidents (capped at $10.9 billion) (Sovacool, 2008a, p206). These subsidies are in addition to numerous 'other' benefits the nuclear industry enjoys: free off-site security, no substantive public participation or judicial review of licensing, and payments to operators to store waste. The existing subsidy established by the Price Anderson Act for nuclear power is estimated to be worth more than the entire US Department of Energy R&D budget for most of the 1990s. A 2009 study from an interdisciplinary team at the University of California Berkeley estimates that at least 37 per cent of the cost for new nuclear units in the US will be completely subsidized by taxpayer dollars (Levin et al, in preparation).

The current obsession with fossil fuel and nuclear subsidies around the world has serious consequences for renewable energy. Such subsidies artificially lower the cost of producing the dirtiest forms of electricity, muddle the market signals that consumers receive, encourage the overconsumption of resources and thus higher electricity use, and lead to capacity developments and consumer patterns in excess of true needs. Forcing renewables to compete with nuclear sources in the current market, but without equal subsidization, is like trying to race a bicycle against a Ferrari. Amazingly, some renewable energy resources such as wind and biogas are already much cheaper and more developed than the current generation of nuclear technologies despite the immense momentum towards conventional technologies.

8.3 CULTURAL AND BEHAVIOURAL BARRIERS

A third category of barriers relates to cultural values and behavioural routines. Interviews with families and their decision-making processes show that direct household electricity consumption is usually based on the *non*-decisions of families. Families do not make explicit decisions about electricity and renewable energy. Rather, they engage in activities of their choice (such as watching *The Simpsons* or washing clothes) and consume energy in the process. Almost never do families decide how many kWh of electricity to use, although family decisions about every other commodity from sugar to shoes, and molasses to movie tickets are consciously planned. Thus, electricity bills are the consequences of the family's lifestyle rather than the other way around (Morrison and Gladhart, 1976). Most families do not make conscious decisions about electricity consumption at all.

As a result of this inverted decision making, and inattention to electricity, misinformation and misunderstanding are widespread. Almost half of respondents in a recent survey identified 'coal', 'oil', and 'iron' as 'renewable resources'. The study also found that the number of participants answering that 'solar' and 'trees' were 'renewable resources' dropped between 1999 and 2004 from 61 per cent to 55 per cent (Kentucky Environmental Education Council, 2005). A 2006 survey of American electricity consumers found that four-fifths were unable to name a single source of renewable energy, even including 'dams' and 'hydroelectric' generators (Shelton, 2006). A 2008 nationwide US Department of Energy study reported that only about 12 per cent of Americans could pass a 'basic' electricity-literacy test. In Europe, a survey of attitudes towards alternative forms of energy in 2007 found that 31 per cent of people in the UK had never heard of wind energy and 90 per cent had never heard of biomass energy in the past year (Reiner et al, 2007). The list could go on, but the point has hopefully been made that many people know little to nothing about electricity, and are therefore more likely to be unconcerned towards renewable energy and FITs (or even hostile, if they've been persuaded by fossil fuel and nuclear industry advertisements).

Strong, albeit subtle, psychological factors and values encourage wasteful electricity consumption and engender opposition towards renewable resources. It is easy to forget that we don't consume energy fuels directly, but instead the services that they provide. With perhaps the odd exception of the drunken fraternity boy consuming gasoline because he was challenged to at a party, or the inmate on death row about to be electrocuted, no one drinks gasoline or channels electricity through their body. We instead seek the mobility, speed, comfort, lighting, cooling, heating, entertainment and enjoyment that those fuels enable.

The problem is that routines and values can be shaped to be unsustainable, and people can come to prefer the existing energy system not because it is optimal, but because it is convenient. Indeed, values relating to comfort, freedom, control, trust, social status, ritual and habit can all shape attitudes for and against renewable energy. One longitudinal study of attitudes towards energy and electricity found that, contrary to expected primacy of concerns about cost, 'comfort' was the single most important determinant of their energy use, an attitude so consistent that neither the location of those surveyed nor the year they were contacted changed the answer (Becker et al, 1981). In other words, the most inelastic and consistent component of electricity consumption, one that did not change with availability of energy, was thermal comfort, making it the most significant predictor of actual energy consumption. The logical extension of this view is that any new technologies that threaten to disrupt the comfort currently provided by the existing energy system, even if only perceptually, will be fiercely resisted. Freedom and control appear to be almost as significant in influencing energy choices. People will resist energy technologies that impede their freedom or appear to diminish their control, implying that many will be unwilling to make any sacrifice for renewable energy at all (Mazis, 1975; Becker, 1978; Brehm and Brehm, 1981; Thompson 1981).

Once values concerning familiarity, energy and consumption are formed, whatever they may be, they tend to be very difficult to alter, especially when transmitted between generations. Psychologists have found that people tend to rationalize all of the decisions they have made afterwards, emphasizing the positive aspects of chosen alternatives and the negative aspects of un-chosen options. As time goes by, individuals come to increasingly view the selected option as clearly superior to all other alternatives (Brehm, 1956). Moreover, the greater the commitment in terms of cost, effort or irrevocability, the stronger and more permanent the effect. People tend to remember the plausible arguments favouring their own positions and the implausible arguments opposing their positions, serving the need for self-justification rather than objective fact seeking (Stern and Aronson, 1984). When applied to the consumption of electricity, renewables and FITs, these psychological factors mean that people resist change because they are committed to what they have been doing. They justify that inertia by downgrading information that implies that change is needed, and this partially explains the failure of people to install their own solar panels or wind turbines.

8.4 Aesthetic and Environmental Challenges

A final challenge relates to the aesthetic and environmental issues associated with renewable energy technologies. Perhaps the most vociferous environmental concern relates to the death of birds ('avian mortality') and bats ('chiropteran mortality') resulting from collisions with wind turbine blades. Centralized and large-scale utility renewable power plants can require large amounts of land, and when these systems are built in densely forested areas or ecosystems rich in flora and fauna, they can fragment large tracts of habitat. Dams can drastically disrupt the movement of species, change upstream and downstream habitats, and can emit greenhouse gases from vegetation rotting in their reservoirs. People have complained that biomass and biogas plants release foul odours near some neighbourhoods, and they can contribute to traffic congestion when large amounts of fuel must be delivered by trucks. Some geothermal plants can emit small amounts of hydrogen sulphide and CO_2 along with toxic sludge containing sulphur, silica compounds, arsenic and mercury. The life cycle for most solar panels requires the mining and processing of silicon and other materials which can contaminate areas of land when such systems break down or are destroyed, such as during hurricanes and tornados.

While the above paragraph may sound bad to advocates of FITs and renewable energy, the facts should be taken in context. The avian mortality issue should certainly be taken seriously, but several facts make bird deaths unique to older wind sites located near bird migration routes and relying on older wind turbines. Moreover, a recent study of the avian deaths from nuclear, fossil-fuelled and wind energy systems found that wind systems were the *best* for birds. The study estimated that wind farms and nuclear power stations are responsible each for between 0.3 and 0.4 fatalities per gigawatt-hour (GWh) of electricity while fossil-fuelled power stations are responsible for about 5.2 fatalities per GWh. The estimate means that wind farms killed approximately 7000 birds in the US in 2006 but nuclear plants killed about 327,000 and fossil-fuelled power plants 14.5 million (See Table 8.1). Fossil-fuelled power stations appear to pose a much greater threat to avian wildlife than wind and nuclear power technologies. Avian wildlife can perish not only by striking wind turbines but can smash into nuclear power plant cooling structures, transmission and distribution lines, and smokestacks at fossil-fuel-fired power stations. Birds can starve to death in forests ravaged by acid rain, ingest hazardous and fatal doses of mercury, drink contaminated water at uranium mines and mills, or die in large numbers as climate change wreaks havoc on migration routes and degrades habitats (Sovacool, 2009b). To put the avian mortality issue in greater perspective, the absolute number of avian deaths from onshore and offshore turbines is incredibly low compared to other sources. Millions of birds die annually when they strike tall stationary communications towers, get run over by automobiles, or fall victim to stalking cats (see Figure 8.3) (MacKay, 2009).

The land employed for wind turbines, unlike the property needed for a coal plant or nuclear facility, can still be used for farming, ranching, and foresting and many forms of solar technology can be fully integrated into buildings and building facades. No form of hydroelectric generation combusts fuel, meaning they produce little to no air pollution in comparison with fossil fuel plants. Dedicated biomass electrical plants release no net carbon dioxide into the atmosphere (as long as they avoid combusting fossilized fuel) and produce fewer toxic gases. High-yield food crops leach nutrients from the soil, but the cultivation of biomass crops on degraded lands can help stabilize soil quality, improve fertility, reduce erosion and improve ecosystem health. Geothermal plants also have immense air quality benefits. A typical plant using hot water and steam to generate electricity emits about 1 per cent of the sulphur dioxide, less than 1 per cent of the nitrogen oxide, and 5 per cent of the carbon dioxide emitted by a coal-fired power plant of equal size.

What happens when the costs and benefits of renewable electricity systems are added together, and compared to other existing alternatives, may surprise readers. The existing electricity rates and prices that customers see on their bills do not reflect many of the costs and benefits discussed above, nor do they reflect them for fossil fuel and nuclear power stations. Economists refer to such items as 'externalities', or costs and benefits resulting from an activity that do not accrue to the parties involved in the activity. Soldiers refer to such things as 'collateral damage'.

Table 8.1 *Avian mortality for fossil fuel, nuclear, and wind power plants in the US*

Fuel source	Assumptions	Avian mortality (total per year)	Avian mortality (fatalities per GWh)
Wind energy	Based on real-world operating experience of 339 wind turbines comprising six wind farms constituting 274MW of installed capacity. Total avian mortality per year taken by applying 0.269 fatalities per GWh multiplied by the 25,781GWh of wind electricity generated in 2006	7193	0.269
Fossil fuels	Based on real-world operating experience for two coal facilities as well as the indirect damages from mountain top removal coal mining in Appalachia, acid rain pollution on wood thrushes, mercury pollution, and anticipated impacts of climate change. Total avian mortality taken by applying the 5.18 fatalities per GWh multiplied by the 2.8 million GWh of electricity produced by the country's fleet of coal-, natural gas- and oil-fired power stations in 2006	14.5 million	5.18
Nuclear power	Based on real-world operating experience at four nuclear power plants and two uranium mines/mills. Total avian mortality taken by applying the 0.416 fatalities per GWh multiplied by the 787,219GWh of electricity produced by the country's nuclear plants in 2006	327,483	0.416

Fossil fuel and nuclear power plants are the world's second largest users of water, produce millions of tonnes of solid waste, emit mercury, particulate matter and other noxious pollutants into the atmosphere, and cause widespread social inequity. Yet in the current system, they don't have to *pay* for most of this damage. If they did have to fully internalize the costs of transportation, air pollution, water contamination and land use (and, when applicable, damages such as injury and death), coal generation would cost US$0.1914/kWh more; oil and natural gas generation US$0.12/kWh more; nuclear power US$0.111/kWh more. In each

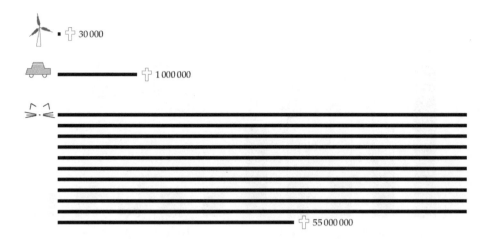

Figure 8.3 *Annual bird deaths in Denmark and Britain caused by wind turbines, cars and cats*

Source: MacKay, 2009

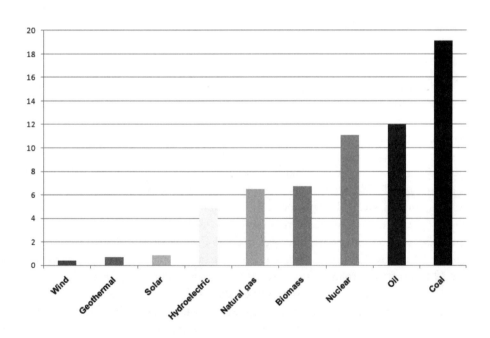

Figure 8.4 *The 'external costs' of conventional, nuclear, and renewable power generators in the US (in 2007 US$/kWh)*

Source: Sovacool, 2008b

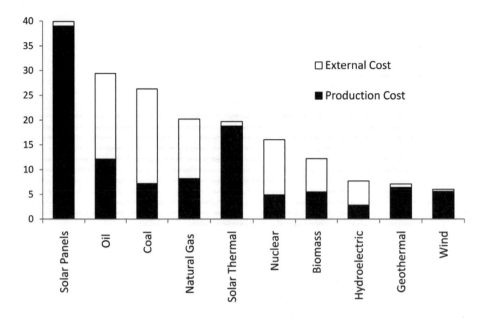

Figure 8.5 *The 'external costs' of power generators compared to their production costs in the US (in 2007 US$/kWh)*

Source: Sovacool, 2008b

case, the external costs for these sources would exceed their current production cost. By contrast, the external costs for renewable energy are much, much smaller: less than US$0.02/kWh for wind, geothermal and solar, about US$0.05/kWh for hydroelectric, and US$0.07/kWh for biomass (see Figures 8.4 and 8.5).

Put in slightly different terms, when correlated with actual electricity generation, the extra costs or externalities associated with conventional sources are monumental: $228 billion in damages per year for coal, $105 billion for oil and gas, $87 billion for nuclear power in the US alone – an amount worth more than the industry's entire revenue for that same year. When roughly quantified and put into monetary terms, the negative externalities for coal power plants are 74 times greater than those for wind farms, and from nuclear power plants negative impacts are 12 times greater than solar PV systems (Sovacool, 2008a, pp113–121).

Clearly – and incontrovertibly – renewables are orders of magnitude better for the environment than any conventional power plant. Renewables do have their own environmental issues, but these pale in comparison to the environmental risks from fossil fuels and nuclear electricity.

Why, then, do some consumers and even environmentalists reject technologies that would bring immense comparative environmental benefits? Given the highly politicized nature of the energy industry, many attempts to 'internalize' the negative external effects of fossil or nuclear energy have failed. Most straightforward would be the introduction of a tax on conventionally produced electricity, although such attempts have been fiercely resisted by utilities and even customers, who don't want to see their rates go up.

Perhaps it is also because renewable electricity systems possess symbolic meaning. Opposition to the siting of new power plants can occur because such technologies inflame pre-existing social conflicts that have little (and sometimes nothing) to do with electricity. Rural residents, for example, often resent urban developers who wish to build electricity projects in their midst. Others oppose new generators because they feel that they have been excluded from the policy making, permitting or siting process. In other cases, rural residents want renewable electricity projects for their own use, as a vehicle for economic development, and resent what seems like meddling by urban residents intent on preserving the countryside for its scenic and recreational value. In this way, renewable electricity technologies become more than simply an electricity generator: they symbolize a method of organizing the landscape, a system of ownership and control, and a personal ethic or a reflection of attitudes (Pasqualetti et al, 2002). One researcher even jokingly commented that modern resistance toward any energy project is so strong that 'not in my backyard,' or NIMBY, is rapidly turning into 'build absolutely nothing anywhere near anything', or BANANA.

Much of this conflict has to do with the immobility of renewable resources. Wind moves but windy locations do not. Wind and sunlight differ from coal and conventional fuels because they cannot be extracted and transported for use at a distant site. For wind farms to be successful, turbines can only be installed where sufficient wind resources exist. Thus, the site-specific nature of wind invites conflict with existing or planned land uses. The landscape itself can shape public attitudes toward renewables, as some landscapes are more valued than others. Place turbines in sensitive areas, perhaps along the coast or in a national park, and prepare for social uproar. Place them out of view or in low-value areas such as sanitary landfills, and opposition diminishes (Pasqualetti, 2004).

Some of the conflict, too, has to do with the misinformation discussed in this subsection. People do not properly assess the costs and benefits of renewable power systems. As only one example, a study of the controversy over erecting an offshore wind farm called Cape Wind near Nantucket Sound, Massachusetts, found that opponents greatly underestimated the environmental value of offshore wind farms. A 2007 survey of 500 local residents living there found that 72 per cent felt the project would have negative impacts on aesthetics and that 42 per cent strongly opposed the Cape Wind project for environmental reasons, even though it would displace about 1.5GWh of more polluting fossil-fuelled capacity (Firestone and Kempton, 2007).

8.5 Conclusion

What are we to make of this mishmash of interconnected impediments?

First, the situation with renewables reminds us that economic signals are necessary but insufficient alone to facilitate consumer acceptance. Utility managers, systems operators, business leaders and ordinary consumers do not function like automatons that rationally input price signals and change their behaviour to optimize benefits and minimize costs. Instead, they are embroiled in a complicated social and cultural environment that is shaped by and helps to shape technological change, rituals, behaviours, values, attitudes, emotions and interests. The impediments to renewable power demonstrate that consumers may reject what can even be in their interest or the interest of society due to lack of information, lack of capital, or sheer selfishness.

Second, the barriers and challenges discussed here provide strong, logical reasons for employing FITs. Government intervention is absolutely essential to correct the market failures, political inconsistencies, cultural bias, and misconstrued environmental impacts that have so far plagued renewable energy deployment. The existing energy market has been invariably shaped by past interventions, and much like the river in the time of Heraclitus, it has been irreversibly altered. Robust public policy is needed to correct the momentum given to fossil fuels and nuclear systems, and create a more just and equitable electricity sector. The market as it exists is incapable of promoting renewables without it.

References

Akerlof, G. A. (1970) 'The market for "lemons": Quality uncertainty and the market mechanism', *The Quarterly Journal of Economics,* vol 84, no 3, pp488–500

Alderfer, B. and Starrs, T. J. (2000) *Making Connections: Case Studies of Interconnection Barriers and Their Impact on Distributed Power Projects*, NREL/SR-200-28053, National Renewable Energy Laboratory, Golden, CO

Allen, A. (2002) 'The legal impediments to distributed generation', *Energy Law Journal,* vol 23, pp505–523

Anderson, D. (1995) 'Roundtable on energy efficiency and the economist – an assessment', *Annual Review of Energy and Environment,* vol 20, pp562–573

Barbose, G., Wiser, R. and Bolinger, M. (2006) *Supporting Photovoltaics in Market-Rate Residential New Construction: A Summary of Programmatic Experience to Date and Lessons Learned*, Lawrence Berkeley National Laboratory, Berkeley, CA

Becker, L. J. (1978) 'Joint effect of feedback and goal setting on performance: A field study of residential energy conservation', *Journal of Applied Psychology,* vol 63, no 4, pp428–433

Becker, L. J., Seligman, C. and Darley, J. M. (1979) *Psychological Strategies to Reduce Energy Consumption*, Center for Energy and Environmental Studies, Princeton, NJ

Becker, L. J., Seligman, C., Fazio, R. H. and Darley, J. M. (1981) 'Relating attitudes to residential energy use', *Environment and Behavior,* vol 13, no 5, pp590–609

Brehm, J. W. (1956) 'Postdecision challenges in the desirability of alternatives', *Journal of Abnormal and Social Psychology*, vol 52, pp384–389

Brehm, S. S. and Brehm, J. W. (1981) *Psychological Reactance: A Theory of Freedom and Control*, Academic Press, New York, NY

Brown, M. A. (1993) 'The effectiveness of codes and marketing in promoting energy-efficient home construction', *Energy Policy*, pp391–402

Brown, M. A. (1997) 'Energy-efficient buildings: Does the marketplace work?', *Proceedings of the Annual Illinois Energy Conference*, vol 24, pp233–255

Brown, M. A, Chandler, J., Lapsa, M. V. and Sovacool, B. K. (2008) *Carbon Lock-in: Barriers to the Deployment of Climate Change Mitigation Technologies*, ORNL/TM-2007/124, Oak Ridge National Laboratory, Oak Ridge, TN

Bulat, O., Danford, L. and Samarasinghe, L. (2008) *Project Sunshine: An Overview of the US Residential Solar Energy Market*, Harvard Business School, Cambridge, MA

Coenraads, R., Voogt, M. and Morotz, A. (2006) *Analysis of Barriers for the Development of Electricity Generation from Renewable Energy Sources in the EU-25*, OPTRES, Utrecht

Coenraads, R., Reece, G., Voogt, M., Ragwitz, M., Resch, G., Faber, T., Haas, R., Konstantinaviciute, I., Krivosik, J. and Chadim, T. (2008) *Progress: Promotion and Growth of Renewable Energy Sources and Systems*, Ecofys, Fraunhofer Institute, Energy Economics Group, LEI, and SEVEn; Utrecht

DeCicco, J., Diamond, R., Nolden, S. and Wilson, T. (1996) *Improving Energy Efficiency in Apartment Buildings*, ACEEE, Washington, DC

Firestone, J. and Kempton, W. (2007) 'Public opinion about large offshore wind power: Underlying factors', *Energy Policy*, vol 35, pp1584–1598

Howarth, R. B., Haddad, B. M. and Paton, B. (2000) 'The economics of energy efficiency: Insights from voluntary participation programs', *Energy Policy*, vol 28, pp477–486

International Energy Agency (2004) *Renewable Energy RD&D Priorities: Insights from IEA Technology Programs*, OECD, Paris, p54

International Energy Agency (2008) *Deploying Renewables: Principles for Effective Policies*, OECD, Paris

Jacobson, M. Z. and Masters, G. M. (2001) 'Letters and responses: The real cost of wind energy', *Science*, vol 294, no 5544, pp1000–1003

Kentucky Environmental Education Council (2005) *The 2004 Survey of Kentuckians' Environmental Knowledge, Attitudes and Behaviors*, Kentucky Environmental Education Council, Frankfurt, KY

Koomey, J. G. (1990) *Energy Efficiency in New Office Buildings: An Investigation of Market Failures and Corrective Polices*, University of California Berkeley, Doctoral Dissertation, Berkeley, CA, p2

Levin, J. E., Hoffman, I. M. and Kammen, D. M. (in preparation) *Costs and Challenges of Significantly Reducing US Carbon Emissions by Expanding Nuclear Power*, unpublished manuscript

Lund, P. (2006) 'Market penetration rates of new energy technologies', *Energy Policy*, vol 34, pp3317–3326

Lund, P. (2007) 'Effectiveness of policy measures in transforming the energy system', *Energy Policy*, vol 35, pp627–639

MacKay, D. (2009) *Sustainable Energy – Without the Hot Air*, UIT Cambridge Ltd, Cambridge, UK, p64

Mazis, M. B. (1975) 'Antipollution measures and psychological reactance theory: A field experiment', *Journal of Personality and Social Psychology*, vol 31, no 4, pp654–660

Mendonça, M., Lacey, S. and Hvelplund, F. (2009) 'Stability, participation and transparency in renewable energy policy: Lessons from Denmark and the United States', *Policy and Society*, vol 27, pp379–398

Meyer, N. (2007) 'Learning from wind energy policy in the EU: Lessons from Denmark, Sweden and Spain', *European Environment*, vol 17, no 5, pp347–362

Morrison, B. M. and Gladhart, P. (1976) 'Energy and families: The crisis and response', *Journal of Home Economics*, vol 68, no 1, pp15–18

Nakarado, G. L. (1996) 'A marketing orientation is the key to a sustainable energy future', *Energy Policy*, vol 24, no 2, p188

Pasqueletti, M. J. (2004) 'Wind power: Obstacles and opportunities', *Environment*, vol 46, no 7, pp22–31

Pasqualetti, M. J., Gipe, P. and Righter, R. W. (2002) *Wind Power in View: Energy Landscapes in a Crowded World*, Academic Press, New York, NY

Pitt, D. (2008) *Taking the Red Tape out of Green Power*, Network for New Energy Choices, New York, NY

Reddy, S. and Painuly, J. P. (2004) 'Diffusion of renewable energy technologies – barriers and stakeholders' perspectives', *Renewable Energy*, vol 29, pp1431–1447

Reiner, D., Curry, T., de Figueiredo, M., Herzog, H., Ansolabehere, S., Itaoka, K., Akai, M., Johnsson, F. and Odenberger, M. (2007) *An International Comparison of Public Attitudes towards Carbon Capture and Storage Technologies*, MIT, Cambridge, MA, p3

Shelton, S. C. (2006) *The Consumer Pulse Survey on Energy Conservation*, Shelton Group, Knoxville, TN

Southern States Energy Board (2003) *Distributed Generation in the Southern States: Final Report*, Energy Resources International, Knoxville, TN

Sovacool, B. K. (2008a) *'The Dirty Energy Dilemma: What's Blocking Clean Power in the United States'*, Praeger, Westport, CN

Sovacool, B. K. (2008b) 'Renewable energy: Economically sound, politically difficult', *Electricity Journal*, vol 21, no 5, June, pp18–29

Sovacool, B. K. (2009a) 'The importance of comprehensiveness in renewable electricity and energy efficiency policy', *Energy Policy*, vol 37, no 4, pp1529–1541

Sovacool, B. K. (2009b) 'Contextualizing avian mortality: A preliminary appraisal of bird and bat fatalities from wind, fossil-fuel, and nuclear electricity', *Energy Policy*, vol 37, no 6, pp2241–2248

Stavros, R. (1999) 'Distributed generation: Last big battle for state regulators?', *Public Utilities Fortnightly*, vol 137 pp34–43

Stern, P. C. and Aronson, E. (1984) *Energy Use: The Human Dimension*, Freeman and Company, New York, NY, pp68–89

Stigler, G. J. (1961) 'The economics of information', *The Journal of Political Economy*, vol 69, no 3, pp213–225

Thompson, S. C. (1981) 'Will it hurt less if I control it? A complex answer to a simple question', *Psychological Bulletin*, vol 90, no 1, pp89–101

Yakubovich, V., Granovetter, M. and McGuire, P. (2005) 'Electric charges: The social construction of rate systems', *Theory and Society*, vol 34, pp579–612

9

Other Support Schemes

In ancient Greek philosophy, the concept of *aporia* indicates a state of perpetual puzzlement.[1] One example comes from Plato, when he has Socrates discuss the essence of knowledge in a dialogue called *Meno*. In this dialogue, the character Meno is left confused over the many possible interpretations of virtue. Plato emphasizes that making a decision between two notions of virtue is easy, but that having to decide between dozens of interpretations is exceedingly difficult. The implicit lesson of the dialogue is that the presence of too much of something, of too many alternatives, can provoke confusion and inaction.

Perhaps proponents of renewable energy should spend more time reading Plato, for they, too, seem to have created too many alternatives when it comes to promoting renewable energy, and they, too, may be responsible for engendering states of *aporia* in their wake. As of early 2009, policy targets for renewable energy existed in 73 countries worldwide including all 27 members of the EU, 33 states in the US and 9 Canadian Provinces. No fewer than 64 countries had energy policies requiring the mandatory use of renewable energy (REN21, 2009). At least 37 countries had adopted some form of price-based support mechanism, 49 states and countries had portfolio standards, and more than 30 countries had tax credits, tax exemptions and public financing (REN21, 2008, 2009). One recent study noted that 55 different types of policy mechanisms were currently in use for supporting renewables around the world (Tonn et al, 2009). Another study, after surveying hundreds of experts in Asia, Europe and North America, identified 30 favoured mechanisms (Sovacool, 2009). Yet another investigation of renewable energy subsidies found that policies can take the form of direct financial transfer (grants), preferential tax treatment (tax credits, exemptions, accelerated depreciation, and rebates), trade restrictions (quotas), financing (low-interest loans), and direct investment in energy infrastructure, research and development (Menz and Vachon, 2006). Feed-in tariffs (FITs) were not included in this last study because they were not considered a subsidy.

This essentially bewildering collection of policy mechanisms leaves a very simple question unanswered: which one, really, is the best? Based on an analysis of the peer-reviewed literature supplemented with extensive discussions and research interviews with experts, this chapter argues that FITs are indeed the best single tool that governments can use to promote renewable energy. The chapter explores the strengths and weaknesses of eight commonly used alternative policy mechanisms for harnessing the power of renewables:

1   renewable portfolio standards and quotas;
2   tradable certificates and Guarantees of Origin;
3   voluntary green power programmes;
4   net metering;
5   public research and development expenditures;
6   system benefits charges;
7   tax credits; and
8   tendering.

The chapter also scrutinizes why empirical evidence (and perhaps common sense) shows that FITs ultimately have advantages over each of them.

9.1 Renewable Portfolio Standards and Quota Systems

Renewable portfolio standards (RPS) set quotas that force suppliers to utilize renewable energy resources.[2] Put another way, RPS, sometimes called 'renewable energy standards', 'sustainable energy portfolio standards', the 'Mandatory Renewable Energy Target' in Australia, the 'Renewables Obligation' in the UK, or the 'Special Measures Law' in Japan, are mandates for utilities to source a specific amount of their electricity sales, or generating capacity, from wind, solar, biomass, hydro and geothermal power plants (Sovacool, 2008a, pp241–261; Sovacool and Cooper, 2007–2008; Sovacool and Cooper, 2008). The RPS in the state of California, for example, currently requires that major electric utilities meet 20 per cent of their retail sales with renewables by 2010 and have a goal of reaching 33 per cent by 2020. In contrast to many RPS schemes in the US, quota schemes in Europe are based on certificate trade for target compliance. Therefore, they are generally called 'tradable green certificate schemes,' and are discussed more in the following section on renewable energy credits.

RPS policies initially developed in the mid-1990s in response to the perceived dangers of the introduction of electricity restructuring, where regulators privatized and liberalized many state electricity markets, and its possible effects on market competition. Since renewable energy sources were not at that time price competitive with a market that did not include the full social costs of electricity, it was agreed that additional policies were required to monetize their positive benefits. RPS became a tool for encouraging renewable energy as support for other mechanisms, such as R&D expenditures and tax credits, were waning (Rickerson and Grace, 2007). In the mid-1990s, the European Commission tried to push member states

Table 9.1 *Countries, provinces and states with renewable portfolio standards and quota schemes*

Year	Cumulative number	Location
1983	1	Iowa (USA)
1994	2	Minnesota (USA)
1996	3	Arizona (USA)
1997	6	Maine, Massachusetts, Nevada (USA)
1998	9	Connecticut, Pennsylvania, Wisconsin (USA)
1999	12	New Jersey, Texas (USA); Italy
2000	13	New Mexico (USA)
2001	15	Flanders (Belgium); Australia
2002	18	California (USA); Wallonia (Belgium); UK
2003	19	Japan; Sweden; Maharashtra (India)
2004	34	Colorado, Hawaii, Maryland, New York, Rhode Island (USA); Nova Scotia, Ontario, Prince Edward Island (Canada); Andhra Pradesh, Karnataka, Madhya Pradesh, Orissa (India); Poland
2005	38	District of Columbia, Delaware, Montana (USA); Gujarat (India)
2006	39	Washington State (USA)
2007	44	Illinois, New Hampshire, North Carolina, Oregon (USA); China
2008	49	Michigan, Missouri, Ohio (USA); Chile; India

Source: REN21, 2009

of the EU to implement quota-based support mechanisms and openly favoured them over price-based support instruments, such as FITs. Today, such quota schemes policies exist in more than 30 states in the US as well as countries such as Australia, Belgium, Canada, Chile, China, India, Italy, Japan, Poland, Taiwan and the UK (see Table 9.1).

Quota schemes possess at least five advantages:

1 While the price of renewable energy will remain uncertain, quotas can ensure a given quantity of renewable energy is delivered by a given date (presuming that such policies are enforced). One study projected that so far the Californian RPS has incentivized $1 billion in new investments in wind energy as utilities rush to make the quota. These include about 3GW of new renewable energy capacity among California's three largest investor-owned utilities and 25,000GWh of renewable electricity as a whole in 2007 (Baratoff et al, 2007; California Public Utilities Commission, 2007).
2 Quota schemes are usually coupled with certificate trading schemes (called RECs or GO schemes, discussed in the next section) giving utilities and states an immense amount of flexibility in meeting targets. Power providers can generate their own renewable energy to meet the target, import renewable electricity from another state, or purchase RECs on the commercial market. Utilities can also decide which eligible renewables they want to deploy and where; no technology is forced upon them.

3 Most quota schemes policies do not require utilities to meet the standard all at once. The policies are gradually phased in over time. An ultimate target of 15 per cent by 2015, for example, may start by saying 11 per cent by 2009, 12 per cent by 2011, and 14 per cent by 2014.
4 The most successful quota schemes are mandatory and have strict penalties for non-compliance, meaning that utilities actually try their best to meet their targets.
5 Quota schemes tend to have relatively low administrative costs and burdens compared to public funding of research and subsidies. Unlike research and development expenditures or tax credits, a quota costs governments next to nothing, since costs are spread among utilities and electricity customers. Since it uses the market as a mechanism to determine the efficacy of any given technology, higher costs (if they occur) are distributed evenly throughout society to those that benefit from them, and blended with the lower costs of existing, conventional generation (Jaccard, 2004). Unlike instruments developed by public utility commissions with long and complex procedures (often followed by litigation), quotas are bureaucratically simple (Lauber, 2004). They avoid the administration or dissemination of funds by government agencies, as quotas enable customers to pay producers directly for renewable energy. And, unlike a one-time award of funds, no project is guaranteed a place in the market (Rader and Hempling, 2001).

While these five advantages do make quota-based systems a superior tool compared to many other options, they do have at least six inherent drawbacks:

1 Quotas stipulate a certain *quantity* of renewable energy but not a certain *price*. Prices are not known up front, and are instead set by the existing market. This means that the price paid to renewable energy suppliers will change as market conditions change; prices will always fluctuate and will never be fixed or predictable (Gipe, 2006). Given that many investors will care more about price than quantity, such an impediment could be significant in the eyes of many banks, insurance companies, and investing firms. Besides, experience in Europe has shown that penalties for non-compliance were often too low, sometimes even below the generation costs of many renewable energy technologies. In this case, utilities would rather pay the penalty than invest in more expensive renewables.

Experience in the UK illustrates the dangers of fluctuating prices quite clearly. There, suppliers can trade Renewables Obligation Certificates (ROCs) to meet quota obligations, and ROC prices are determined by the market and are therefore constantly changing, so developers are charged more for borrowing capital and have to charge more for the power they produce.[3] Professor David Elliott from the Open University believes that this explains why the same type of renewable energy project can cost up to twice as much in the UK compared to other places such as Denmark and Germany.[4] Similar observations were made by Dinica (2006).

2   Quota-based systems in general have been implemented very inconsistently, with a shocking number of different requirements. Inconsistencies between RPS programmes over what counts as renewable energy, when it has to come online, how large it has to be, where it must be delivered, and how it may be traded clog the renewable energy market in the US. Implementing agencies and stakeholders must grapple with inconsistent RPS goals, and investors must interpret competing and often arbitrary statutes. Some RPS policies even include fossil fuel energy systems such as natural gas fuel cells, combined heat and power, and clean coal as 'renewables' (Sovacool and Cooper, 2007, p50). Such inconsistency stifles investment in the renewable energy market. One study of the financial community in the US concluded that instability and lack of consistency in state RPS markets created too much uncertainty for investors and led to considerable delays in starting interstate projects that spanned more than one RPS programme (Baratoff et al, 2007).

3   Because quota policies are a least-cost mechanism, they are less flexible in offering targeted support for specific renewable energy technologies and less effective at ensuring diversification among a broad class of renewable energy systems. Some commentators used the term 'dynamic inefficiency' to refer to policy mechanisms that fail to promote a basket of technological alternatives (Ragwitz et al, 2009). Under the intense competition induced by an RPS to keep prices low, suppliers tend not to support higher cost renewable energy resources such as small-scale wind or solar (Center for Resource Solutions, 2001). To date, the bulk of projects associated with RPS programmes have been commercial-scale wind farms. This means RPS tend to favour vertically integrated generating companies and big electric utilities that can handle large-scale renewable energy power plants. One recent study from the US National Renewable Energy Laboratory on the effectiveness of policy mechanisms argued that RPS have been completely *ineffective* at promoting small-scale, residential renewable energy systems (Coughlin and Cory, 2009).

4   To ensure flexibility many quota schemes rely on tradable certificates, but this only opens them up to the litany of problems associated with tradable credits, discussed in the next section. These problems significantly add to the cost of renewable energy projects under an RPS. One study looking at estimates of renewable energy supply to meet 28 of the RPS policies in the US found that development will be uneven across the country. Some areas will soon have a deficit of supply while others will have a surplus. The study projected regional shortages of renewable energy capacity in New England, New York, and the Mid-Atlantic states and surpluses in the Midwest, Heartland, Texas, and the West (Bird et al, 2009). The problem is that such shortfalls will require tradable credits and inter-regional transfers of renewable power where transmission is available to meet compliance. Both of these mechanisms will add considerably

to the cost of renewable electricity; credit trading brings with it all of the transaction costs discussed below and inter-regional transfers have higher transmission and distribution efficiency losses.

5   Quotas appear to be ineffective at promoting sustained, rapid growth in renewable energy, at least in the US and the UK. One assessment estimated that RPS programmes have been responsible for only one-fifth of renewables growth in the US from 1978 to 2006, with the remaining growth coming from other mechanisms (see Figure 9.1). In the UK, the Renewables Obligation has failed to grow the renewable energy market as intended and has actually increased costs. The uncertainty of prices for renewable electricity under the scheme has passed regulatory risk to the private sector, which then place a premium on it. The uncertainty of price causes leakage away from developers and encourages suppliers to take a margin to deal with the risk, meaning less capital is available for actual power projects (Carbon Trust, 2007).

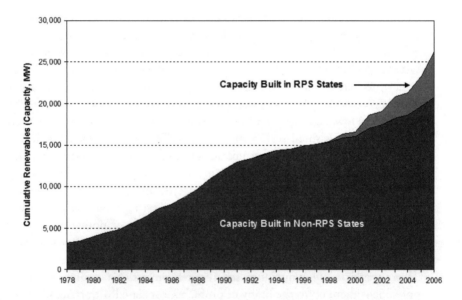

Figure 9.1 *Renewable energy capacity built in states with and without renewable portfolio standards (RPS) in the US, 1978–2006*

Source: Wiser, 2007

6   Quantity-based support mechanisms such as quotas can limit the expansion of renewable energy support. If the quota is based on certificate trade, producers will try not to achieve the set target, as otherwise the value of their certificates will fall to zero. Quota schemes therefore can be prone to strategic gambling. In oligopolistic markets, actors will try not to reach the target in order to keep

the certificate price artificially high. In Section 4.8 we have argued against 'caps' in the case of feed-in tariffs. As capping capacity is an immanent part of a quota scheme, we cannot recommend this support mechanism.

For perhaps these reasons, many electric utilities and energy suppliers are failing to meet their actual quotas. Even when all RPS programmes are included in projections, the contribution of renewable resources is unlikely to exceed 3 per cent of total electricity supply by 2015 and 4 per cent by 2030 in the US (Sovacool, 2008b). Assuming full compliance is achieved for every RPS programme, 60GW will be added by 2025 in the US, equivalent to only 4.7 per cent of projected generation and only 15 per cent of projected growth in electricity demand (Wiser and Barbose, 2008).

Even this amount of capacity is unlikely, given that several states are struggling to meet their RPS targets. More than $18 million in alternative compliance payments were paid in 2006 in Arizona (where RPS compliance was below 50 per cent from 2003 to 2007) and Massachusetts (where the difficult political climate has stalled the offshore Cape Wind project). Nevada has consistently struggled to sign contracts for RPS projects, and New York missed its first year target by a wide margin (Wiser and Barbose, 2008). As the US Department of Energy put it, even in a world with state RPS programmes 'oil, coal, and natural gas still are projected to provide roughly the same 86 per cent share of the total US primary energy supply in 2030 that they did in 2005' (US Energy Information Administration, 2007).

9.2 Tradable Certificates and Guarantees of Origin

Renewable energy credits (RECs), sometimes called 'green tags' and 'tradable green certificates' (TGCs) in Europe and 'Renewables Obligation Certificates' (ROCs) in the UK, are certificates that authenticate that 1MWh of electricity was generated from a qualifying renewable resource. When we are talking about RECs in this chapter, it should not be confused with the European Renewable Energy Certificate system, which tends to be used for electricity disclosure and marketing strategies. The major difference between RECs, TGCs or ROCs, Guarantees of Origin (GOs) and the European Renewable Energy Certificates is that the former are traded for target compliance under national, obligatory support schemes while the latter are usually used for electricity disclosure for voluntary green electricity purchase. Still, some key commonalities exist. Each certificate usually lists a comprehensive amount of information, including the location and type of facility generating renewable electricity, its operator, the company owning it, the time and date generated, and sometimes pollution and emissions of greenhouse gases displaced. The idea is that RECs represent commodities that can be traded on the market, with those purchasing RECs able to say they bought one hour of cleaner electricity.

To make matters a little confusing, this isn't always true. RECs can be bundled or unbundled. Bundled RECs mean that the physical electricity *plus* the certificate were transferred together. Say a company in Los Angeles wanted to prove it liked renewable energy, needed electricity, and was connected to the grid. It could then purchase 1 MWh of electricity from a local renewable power provider and also get one REC to affirm that it was indeed from renewables sources, keep this REC to satisfy a possible regulatory requirement or sell it on the 'secondary market' to a voluntary green power programme.

Unbundled RECs refer to the certificate without the electricity. They are based on the idea of 'physical detachment' or 'virtual trade', the notion that credits can be traded independently of the actual electricity associated with them, and therefore traded between utilities and governments to prove compliance with different renewable energy goals and targets (Ragwitz et al, 2009). Say a company was located in an area without renewable power suppliers, or that the transmission constraints of receiving renewable electricity from far away were too great, or that the company didn't actually need electricity. That company could then purchase an unbundled REC and still claim they supported the production of renewable electricity, even if they didn't really use it.

Sometimes, regulations stipulate only that renewable energy has to be generated from within a certain community, county, state or country. In these circumstances, the RECs will tend to include only information about when and where it was generated. These types of RECs are often called Guarantees of Origin (GOs), and require countries with renewables quotas to certify that renewable electricity actually comes from where it is supposed to, much like a label on an article of clothing tells consumers where it was made. A regulatory body such as a transmission operator or public utility commission is granted the authority to certify such guarantees.

RECs are thus more detailed and comprehensive than GOs, although the latter are almost always bundled with the electricity they generate, and intended to travel along with that electricity across the border (*Refocus*, 2003). At least 21 REC schemes were up and running around the world in 2008, and GOs are widely used in the EU, where the Renewable Electricity Directive requires member states to produce electricity from domestic renewable resources. Guarantees of Origin are currently operational in 16 EU member states. Norway, Switzerland and Iceland have their own systems and three separate systems are used in Belgium (Coenraads et al, 2008). These systems issue actual guarantees, usually called Renewable Energy Guarantee of Origin Certificates, or REGOs, in 1MWh blocks as evidence to their customers that renewable power was generated within each system. It must be noted, however, that Guarantees of Origin are so far only used for disclosure purposes, not for target compliance, despite attempts by the European Commission.

RECs and especially REGOs are intended to improve transparency and enhance the information for consumers about the power they purchase. The dilemma is that 'green' electricity looks the same as 'dirty' electricity. As one study put it:

> *The grid can't tell a green electron from a grey one, so the consumer can't see the quality of the power he receives. Green power is more or less virtual and can only be certified in an administrative way. In fact a customer 'greens' their purchased power by buying the guarantee that the supplier has bought exactly the same amount of green electricity from a green generator.* (Langeraar and de Vos, 2003)

RECs and REGOs therefore serve as proof that the supplier actually provided them with a green product. They can function as an accounting system to prevent cheating and ensure compliance.

RECs are also designed to increase the flexibility for electric utilities and power providers having to comply with renewable energy regulations and targets. RECs provide electric utilities and other power suppliers with choices similar to the way emissions control strategies implemented in the 1970s and 1980s worked to reduce lead pollution from refineries and chlorofluorocarbons from aerosols, and in the 1990s lowered nitrogen oxide and sulphur dioxide emissions. Cap-and-trade policies set an upper limit for emissions for a given time period and reduce emissions limits over time. Polluters can either reduce their own pollution or buy certificates that represented emissions reductions beyond mandated targets. In a similar way, REC systems allow utilities and providers to generate their own renewable energy, purchase renewable energy from others, or buy credits. It therefore blends the benefits of a 'command and control' regulatory paradigm with a 'free' market approach to environmental protection. It is often used in conjunction with voluntary green power programmes (discussed below) and renewable portfolio standards and quota systems (discussed above).

Notwithstanding these benefits, REC systems face at least five serious challenges:

1. First, they are expensive and complex. The trading of RECs is time consuming and involves transaction costs on both sides: for those who must produce, certify and sell a REC, and for those who must purchase and verify the authenticity of a REC. Even voluntary programmes suffer from these transaction costs. Surcharges from brokers often occur on both sides of the transaction (one trader unofficially told one of the authors that they usually get a 10 per cent commission on every REC that they sell), and national REC systems usually require a formal registry and auditors that can monitor transactions to avoid double counting.

    The high administrative expenditure associated with REC systems tends to exclude small and medium-sized enterprises from participating, and in some cases hurts competition in the market. Since smaller firms often cannot afford to set up their own trading of RECs, they tend to go through larger utilities for power purchase agreements. Ironically, this can force small firms to reveal all information about their projects to, in essence, their competitors, who can then use that knowledge against them at a later date.[5]

Such complexity (and the transaction costs along with it) is exacerbated by a lack of uniformity between REC trading systems. In the US, state-by-state differences and restrictions have splintered the national REC market into regional and state markets with conflicting rules (Sovacool and Cooper, 2007). In just the Northeast, for example, the electricity wholesale market is controlled by three independent system operators – ISO-NE (New England), NYISO (New York) and PJM (13 mid-Atlantic states). In August 2005, PJM launched its Generation Attribute Tracking System (GATS) to monitor RECs between PJM member states. While GATS will help facilitate a robust REC trading market between PJM members, its convoluted rules hamper REC trading outside of its geographically defined service area. Generators external to PJM are allowed to trade RECs in the GATS market, but must qualify for one of the RPS policies of a PJM member state and must be physically located adjacent to PJM geographical boundaries. However, some PJM member states (Delaware, Maryland and Washington DC) impose an additional requirement that the electricity from renewable generators outside of PJM be imported into the territory in order for external generators to freely trade RECs within their states (Sovacool and Cooper, 2007). PJM member-states also differ conspicuously in their treatment of RECs from generators within the service area. In Maryland and Pennsylvania, generators are allowed to bank their RECs for up to two years after the year of generation. But in Rhode Island, generators may only bank up to 30 per cent of their compliance total (and then only if the banked RECs are in excess of the compliance total in the year of generation).

Contributing to the complexity, ISO-NE has its own REC trading market supported by the Generation Information System (GIS). GIS sets stringent limits on who can trade within the ISO-NE region, regardless of the individual state RPS policies. GIS also requires that generators operate in control areas that are directly adjacent to ISO-NE, further distorting the REC trading market. Generators in NYISO, for example, can trade RECs in Massachusetts, but generators in PJM cannot. Connecticut further restricts REC trading to generators actually within ISO-NE, but, to complicate matters even more, that restriction may expire in 2010 (Sovacool and Cooper, 2007).

In the EU, trading systems for REGOs are just as convoluted. Denmark has a central registry, Romania does not. Austria allows guarantees to be transferable, France does not. Estonia standardizes them according to European Commission directives, Poland does not (Coenraads et al, 2008).

2 RECs force investors to deal with uncertain prices. REC prices have been especially volatile in the US. The inconsistencies between REC definitions and their compliance mechanisms have caused spot REC prices to vary substantially

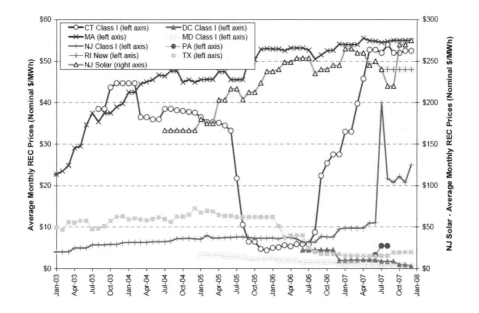

Figure 9.2 *Price volatility of RECs in the US, 2003–2008*

Source: Wiser and Barbose, 2008

across regions and across renewable technologies. Because some states allow out-of-state RECs to apply to in-state mandates, significant price fluctuations are possible even within a single service area. For example, the wholesale price for wind-derived RECs ranged anywhere from $1.75/MWh in California up to $35/MWh in the Northeast in 2006. For biomass RECs, the price ranged from $1.50/MWh in Western states to $45/MWh in New England. For solar-derived RECs, the wholesale prices *in one service area* (WECC) ranged anywhere from $30–150/MWh depending on the state (see Figure 9.2).

Christopher Berendt, of Pace Global Energy Services, has noted how the volatility of REC prices actually *limits* the investment capital available for new renewable energy projects:

> *While state systems share similarities, there is a critical lack of consistent fungibility between RECs issued in different states and control areas ... Thus, there are no real REC 'markets' among or even within the states, only individual state regulatory compliance 'systems.' The lack of a real national REC market ... creates an absence of liquidity for RECs and thus for investment capital.* (Berendt, 2006, p57)

Renewable energy investors require reliable information and predictable rates of return from the start of the financing process. Researchers at the Lawrence Berkeley National Laboratory have tracked the wild fluctuation of REC prices and found them to be a significant deterrent to renewable energy investment. They reported that 'these fluctuating prices have, in some cases, impeded renewable energy development because they offer unclear price signals to renewable energy investors about the attractiveness of development activity' (Wiser et al, 2007b).

3   By unbundling renewable electricity from credits, tradable credit schemes create a de facto segregation of electricity markets. Renewable electricity and the manifold benefits it brings, such as diversification, cleaner air, and better jobs, goes to one community while the credit goes to another. This can lock in existing asymmetries where renewable-resource-rich regions become cleaner and healthier but renewable-resource-poor communities, which end up buying RECs, become worse off. The REC system can become self-replicating because once a region becomes dependent on importing RECs they will usually not have the funds to build their own renewable energy capacity, creating the need to purchase more RECs.

4   Because the major aim of REC systems is to increase flexibility and lower costs, they tend to favour least-cost technologies, not a rich assortment of different (and less mature) renewable energy resources. 'All too often', an anonymous high ranking official in the US confided to one of the authors, 'the flexibility involved with REC trading schemes means only 'flexible to make more money'.[6] Certificate trading can enable some companies to extract windfall profits. While literally thousands of studies critiquing tradable certificate schemes have been published in the past five years, one from 2009 is most telling. The study looked at the performance of national REC programmes in Flanders, Sweden and the UK. The study found that in *each case* REC schemes favoured incumbent companies and large utilities, that they only invested in the cheapest renewable resources (and did not develop less mature technologies), and tended to induce excessive levels of high profits (Jacobsson et al, 2009).

5   REC schemes are prone to a problem known as additionality. That is, they tend to support only renewable energy projects that would have occurred anyway. Given the inconsistency between REC programmes and problems with unpredictable prices, many project developers do not rely on the potential sale of RECs when they decide to invest in a renewable project. Instead, they consider RECs as 'gravy,' something that is nice on top of what they expected but by no means central to their investment (Baratoff et al, 2007).

9.3 VOLUNTARY GREEN POWER PROGRAMMES

Sometimes the trading of credits or purchases of renewable electricity is not mandatory. Voluntary green power programmes, sometimes called 'green power marketing', 'voluntary green power markets', or 'utility green pricing', enable consumers to voluntarily pay more to receive electricity from renewable resources. (When their local power provider is unable to actually deliver this electricity, many programmes allow consumers to buy renewable energy credits, a mechanism explained further below.) The key difference between voluntary green power programmes and mechanisms such as tradable credits, therefore, is that green power programmes can include *both* credits (e.g. based on REGOs) and actual electricity.

Under green power marketing schemes, the virtue of renewable energy is used as a marketing tool to interest consumers. The customer – whether an individual, business or institution – can join a programme offered by a local electric utility or retail marketer to purchase renewable energy (in areas where it is provided) or credits (in areas where it is not) (Birgisson and Peterson, 2006).

Given the absence of national government support for a renewable portfolio standard or FIT, green power programmes have become especially popular in

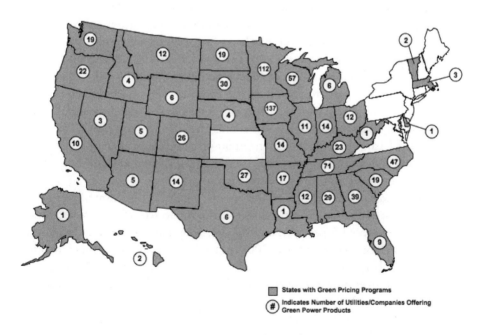

Figure 9.3 *Green power programmes in the US, 2008*

Source: Bird et al, 2009

the US. As of September 2008, more than 850 utilities in 40 states offered some type of green power programme (see Figure 9.3). While the numbers vary based on who does the counting, about 850,000 residential and commercial customers participated in these green power programmes and purchased 18TWh of electricity in 2007. Top municipal buyers of 'green power' included the city of San Diego, Austin Independent School District, and buying groups in Montgomery County, Maryland, New York State, and East Bay Municipal Utility District in California. Top commercial purchases were the US Air Force followed by a list that includes Whole Foods Market, Johnson & Johnson, Starbucks, HSBC North America, University of Pennsylvania and the World Bank Group. In Germany, more than one million households and commercial firms relied on green power programmes to purchase 4.1TWh of electricity; in Australia, almost one million homes and 34,000 businesses purchased 1.8TWh of green power; and in Switzerland more than 600,000 customers purchased 4.7GWh (REN21, 2009, pp20–21).

Green power programmes have two primary strengths. First, they have the advantage of allowing customers in places that do not have significant renewable resources to support the development of renewable energy technologies elsewhere. Second, they do not impose the costs of renewable energy on those that do not wish to pay for them.

These strengths, however, are offset by substantial weaknesses:

1   Green power marketing schemes provide no guarantee that additional renewable energy capacity will be built. The most common experience with green power programmes growing rapidly has been for programme sponsors to cap or limit the programme, not build more capacity. In 2005, for example, Xcel Energy and Oklahoma Gas & Electric quickly and fully subscribed their green power programmes but then had to refuse to let additional customers participate. Similarly, Austin Energy was forced to implement a lottery when its GreenChoice product fell below standard electricity rates (Bird et al, 2008, p5). The lesson seems to be that green power programme managers have little incentive to improve or expand their programmes if they are already receiving a stable revenue stream from customers.
2   Green power programmes rarely represent a significant fraction of energy use or electricity sales. For those programmes run by electric utilities, participation rates rarely exceed 5 per cent, and the most popular programmes have never exceeded 20 per cent (Bird et al, 2008, p1). Green power programmes, in other words, are being used by a very small fraction of customers. In 2008, the top ten largest green power programmes, the ones with the highest participation rates, had only 398,488 customers enrolled (see Table 9.2). This number may sound impressive, but it represents less than 0.5 per cent of the nation's 120 million residential electricity customers. The problem here is that because green power programmes are not mandatory, customers can opt for dirty and

Table 9.2 *Number of utility green power participants for the ten most successful programmes, December 2008*

Rank	Utility	Programme(s)	Participants
1	Xcel Energy	Windsource Renewable Energy Trust	71,571
2	Portland General Electric	Clean Wind Green Source	69,258
3	PacifiCorp	Blue Sky Block Blue Sky Usage Blue Sky Habitat	67,252
4	Sacramento Municipal Utility District	Greenergy	45,992
5	PECO	PECO WIND	36,300
6	National Grid	GreenUp	23,668
7	Energy East (NYSEG/RGE)	Catch the Wind	22,210
8	Puget Sound Energy	Green Power Program	21,509
9	Los Angeles Department of Water & Power	Green Power for a Green LA	21,113
10	We Energies	Energy for Tomorrow	19,615

Source: National Renewable Energy Laboratory, 2009

conventional electricity at cheaper rates and 'free ride' on the environmental benefits provided by those actually subscribing to the programmes (Duke et al, 2005).

3   Green power programmes, because they try to avoid charging consumers too much, tend to promote only the lowest-cost renewable resources. Indeed, the programmes in the US have almost exclusively promoted large-scale wind farms, but not distributed solar panels, small-scale wind turbines, or other alternatives. In Europe, voluntary markets for green power have been primarily based on cheap hydroelectric power mostly produced and certified in Scandinavian countries and sold in central Europe.

4   Ironically, given the point above about keeping costs low, green power programmes *do* tend to be more expensive than other policy mechanisms. This is because the programmes need firms to certify credits, match buyers with sellers, track trades, and ensure no 'double counting' occurs (i.e. that the same credit is not used more than once). Some of these problems are discussed further below when talking about renewable energy credits, but these extra transaction costs do add to the expense of green power programmes. In 2009, for example, the average purchase price for wind electricity from a green power programme in the US was $0.091/kWh (Gomez, 2009) when the US Department of Energy reported that the average cost of producing and transmitting that electricity was less than $0.07/kWh (US Department of Energy, 2008). This implies an extra cost of about $0.02/kWh merely to manage the programme.

Unfortunately, these extra costs mean green power programmes are also the first to be cut during economic downturns. From 2007 to 2008, when the global economy was relatively healthy, local governments and municipalities increased their green power purchases by 200GWh. From 2008 to 2009, in the midst of the global financial crisis, they increased their purchases by only 17GWh. The City of Durango, Colorado, for example, used to buy electricity for all government buildings from green power programmes, but the City Council cancelled the programme in 2009 and reverted to electricity from coal plants to save money (Gomez, 2009).

9.4 Net Metering

Net metering enables owners of grid-connected renewable electricity systems to be credited for the electricity that they provide to the grid – in effect, to spin their meters in reverse. It should be pointed out that unlike many of the other mechanisms discussed in this chapter, net metering does not really 'compete' with FITs and can be used to complement FIT goals.

Classically, whenever customers wanted to actually produce renewable energy at home and then 'sell' it or 'transfer' it to the grid, they had to use two separate meters ('double metering') or complicated calculations. Double metering, however, often gave unfair rates to renewable energy suppliers. The best example comes from the US, where customers wanting to sell power back to the grid in the 1990s would be charged about $0.12–0.19/kWh for the power they consumed, but would get only $0.013/kWh (something called the 'avoided cost') for the power they generated and sold, as little as 7 per cent of the retail cost of power (Bergey Windpower, 1999). The second problem with double metering was that it involved extraordinary transaction costs, often with utility employees travelling directly to homes to verify power generation and spending extensive time writing and processing cheques. During the era of double metering, an owner of one small-scale wind system in Oklahoma City, for example, possessed a collection of more than 60 cheques from Oklahoma Gas & Electric in amounts ranging from $0.02–0.37 for one month (Bergey Windpower, 1999).

Net metering is designed to eliminate these complications, and allow customers to automatically track the power they both consume (from the grid) and produce (to the grid) with a single meter (see Figure 9.4). As of March 2009, net metering was available in 47 states and provinces in North America (Canada and the US) along with national programmes in Belgium, the Czech Republic, Denmark, Italy, Japan, Mexico and Thailand. In theory, net metering has a number of potential benefits. Because some renewable resources produce electricity when it is valued the most, such as a solar panel powering the grid at 2pm on a hot summer day, net metering systems (when coupled with time-of-use rates) can enable them to receive credit for this more expensive power. Net metering reduces administrative costs for utilities and power providers as well, as they no longer have to install two meters or read meters manually.

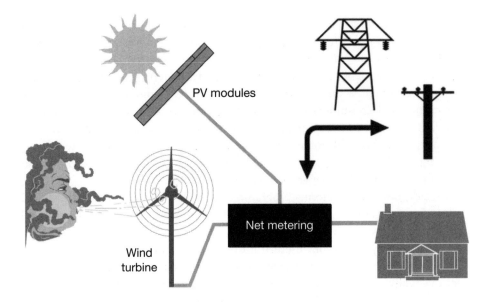

Figure 9.4 *Simplified diagram of net metering showing power flows to/from the grid*

Source: Benjamin K. Sovacool

Perhaps for these reasons, net metering has played a significant role in encouraging investment in distributed renewable energy systems. Under two of the most successful net metering regimes, customers in California and New Jersey had installed more than 23,000 distributed solar systems, collectively, by early 2008. Net metering has been described as 'providing the most significant boost of any policy tool at any level of government … to decentralize and "green" American energy sources' (Network for New Energy Choices, 2007). By compensating customers for reducing demand and sharing excess electricity, net metering programmes are powerful, market-based incentives that states have utilized to promote renewable energy.

Because net metering enables consumers to make more money from selling renewable power, and also empowers them to purchase less electricity from utilities, they have often been fiercely opposed by the electricity industry. Intense pressure from industry groups in the US, for example, has convinced most states to limit the aggregate capacity of eligible net metering customers to a small percentage of a utilities' peak load. Also, in most states, producers are credited only up to the amount of electricity that they consume; any excess beyond the level of consumption goes to the utility.

One recent evaluation of state net metering programmes found that the most successful did not set limits on maximum system capacity or restrictions on eligible renewable resources and tended to require that all utilities participate.

These programmes also usually included all classes of customers, offered consistent interconnection standards and had little to no application fees, special charges or tariffs. However, the study also noted that most programmes were *not* designed that way (Haynes, 2007). In Ontario, Canada, net metering is only permitted for systems up to 500kW and excess power generated (e.g. homes that produce more power than they consume) can only receive 'credits' to offset future consumption, not cash (Ontario Ministry of Energy, 2008). In Idaho and Virginia, regulatory authorities have been able to impose insurance requirements that force homeowners wanting to take advantage of net metering to pay an extra $100,000–300,000 in insurance coverage (US Department of Energy, 2006).

In addition, most metering programmes only allow a 'credit' equivalent to the price of conventional electricity, therefore failing to reflect the full environmental benefits of renewable energy.

And, finally, while net metering does tend to stimulate deployment of distributed renewable energy systems at the residential and commercial scale, it does virtually nothing to promote large renewable energy power plants. In no country has net metering managed to bring about a substantial shift in overall capacity to renewable resources. The explanation may be that the investment security for renewable energy producers is relatively low compared with the fixed rates offered by FITs. As pointed out in Section 2.3, we do not recommend linking the remuneration of renewable energy projects to electricity prices because these prices will fluctuate. Net metering does not reduce or eliminate this form of uncertainty and volatility.

9.5 Research and Development (R&D) Expenditures

Another way many governments promote renewables is by directly paying for research and development (R&D). Every country in the EU and North America manages some sort of R&D programme. In the US, which is one of the largest spenders on energy R&D in absolute terms, more than 150 separate R&D programmes are funded by the federal government. These programmes tend to fall into eight major energy activity areas:

1. energy supply;
2. energy's impact on environment and health;
3. low-income energy consumer assistance;
4. basic energy science research;
5. energy delivery infrastructure;
6. energy conservation;
7. energy assurance and physical security; and
8. energy market competition and education.

The R&D programmes are run by 18 federal agencies, including the Department of Energy, Department of Agriculture, and Department of Health and Human Services (US Government Accountability Office, 2005). Some countries spend more on individual technologies. In 2003 the US government spent $139 million for R&D on photovoltaics, yet that same year Japan spent more than $200 million and Germany more than $750 million (US House of Representatives Committee on Science, 2006).

R&D strategies have the benefit of being extremely flexible. Policy makers can support any particular technology, and can easily control and monitor the distribution of research funds. In addition, the direct expenditures on R&D can create jobs and occasionally lead to patents and royalties owned by the government.

However, there is no guarantee that R&D strategies will pay off, and the large nature of some research projects can encourage excess and corruption. Prominent examples of programmes in the energy sector that continued to receive funding long after they were determined technologically unfeasible include the Clinch River Breeder Reactor (a $2.5 billion demonstration liquid-metal fast breeder reactor plant); a magneto-hydrodynamics programme (a $61 million fossil energy programme attempting to use electromagnetic induction to produce electric power from coal); and the creation of the Synfuels Corporation, a $2.1 billion synthetic fuels programme established in 1981 to develop alternatives to oil (Gallagher et al, 2004).

R&D efforts are also even more susceptible than tax credits to declining government budgets. The budget authority for the US Department of Energy to conduct research and development on renewable, fossil and nuclear generating technologies, for instance, declined by more than 85 per cent (in real terms) between 1978 and 2005, dropping from $5.5 billion to $793 million (US Government Accountability Office, 2006). The same has occurred globally for all R&D expenditures. Direct public expenditures on energy R&D for all IEA countries has fluctuated from $11 billion in 1974 to a high of about $20 billion in 1980, only to drop to $14 billion in 1985 and, after starting to rise in the late 1990s, peak at only $12 billion in 2007 (see Figure 9.5) (Jørgensen, 2006). It has been equally inconsistent for specific expenditures on renewable energy (see Figure 9.6) (IEA, 2008, p157).

9.6 SYSTEM BENEFITS CHARGES

System benefits charges (SBCs) place a small tax on every kWh of electricity generated, and then collect and use those funds to pursue socially beneficial energy projects (Haddad and Jefferiss, 1999, p70; Wiser et al, 2002, p1; Bolinger et al, 2005). Most SBCs are incredibly small. In the US, the most expensive charge is $0.0005/kWh. In the state of Massachusetts, for example, renewable energy

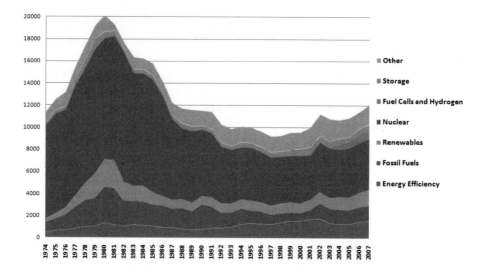

Figure 9.5 *Global energy R&D expenditures, 1974–2007 (millions of US$)*

Note: For IEA member countries only

Source: IEA, 2009

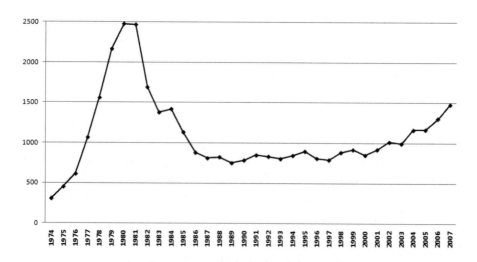

Figure 9.6 *Global energy R&D expenditures on renewable energy, 1974–2007 (Millions of US$)*

Note: For IEA member countries only

Source: IEA, 2009

activities are funded by an SBC of $0.000005/kWh. However, because so many kWh are generated, the SBC still produces millions of dollars of revenue each year.

SBCs, also called public benefit funds, system benefit funds, and clean energy funds, originated in the 1990s along with RPS policies at a time when state policy makers considered electric utility restructuring legislation. Afraid that gains made in pursuing research, development and implementation of environmentally preferable renewable energy technologies would end after regulators lost their sway, advocates of the novel technologies in some states won concessions for a new funding mechanism for high-risk or long-term projects. The basic premise is that without an SBC, socially desirable projects in the energy sector, such as assistance to low-income consumers or investments in renewable energy, will not occur in competitive electricity markets because such projects tend to be less profitable in the short term. SBCs allow such programmes to continue.

First implemented in Washington State in 1994, SBCs were endorsed as a way to fund services that had previously been included in customers' bills of regulated utility companies (Washington Utilities and Transportation Commission, 1994; FERC, 1995). California was quick to follow with an SBC programme that has expended at least $872 million on renewable energy and energy efficiency projects from 1998 to the end of 2001.[7] Today, at least 18 different SBC programmes exist in the US, and other countries have utilized a variety of SBC arrangements. The UK has an 'Energy Savings Trust', operating as a private limited company, that receives money from a charge on transmission and distribution services which it uses to invest in energy efficiency and renewables such as solar water heating and geothermal heat pumps. Norway has a small 'transmission tax' dedicated to funding

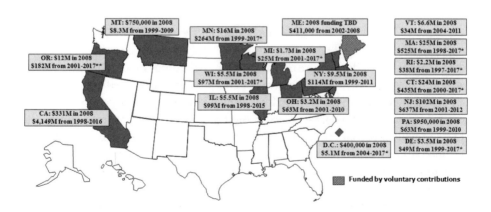

Figure 9.7 *System benefits charges and revenues in the US, March 2008*

Source: Interstate Renewable Energy Council, 2008

clean energy projects. New Zealand created its 'Energy Saver Fund' to support US$18 million worth of investment in small-scale renewable energy and energy efficiency (Bradbrook et al, 2005).

SBCs have at least three benefits:

1 They raise significant amounts of money. The 18 funds in the US alone are expected to generate almost $7 billion in energy investments from 1997 to 2017 (see Figure 9.7).
2 They have the advantage of socializing the cost of renewable energy among all ratepayers in a given region, forcing everyone to pay for the cleaner energy that will benefit them.
3 SBCs offer policy makers the flexibility of promoting energy efficiency and other energy projects besides renewables. One survey of the SBC programmes in the US found that they promoted large-scale wind farms and solar power plants; non-renewable forms of distributed generation; consumer financing for energy efficiency investments; support for green power programmes; public education campaigns about energy; small grants for business development; local research and development on new technologies; and low-income energy assistance and weatherproofing programmes (Wiser, 2007).

SBCs also suffer from drawbacks. Their narrow geographic focus means that, to date, they have been deployed only in single states or on in-state utilities (Wiser et al, 2002, pp130–132). They have been modestly funded compared to expenditures on electricity and the amount of funding differs greatly by state, with only a small fraction of this money going towards renewables. Of the $5 billion that will flow into SBCs from 2006 to 2016, only 10 per cent, or $500 million, will fund renewable energy projects (Wiser, 2007).

9.7 Tax Credits

Two types of tax credits feature prominently in the promotion of renewable energy around the world: investment tax credits and production tax credits.

Investment tax credits (ITCs), like the name implies, give favourable tax treatment to taxpayers who decide to invest their money in renewable energy projects. ITCs provide a partial tax write-off to investors in a particular renewable energy technology. At least 30 countries offered ITCs for renewable energy in 2007. ITCs have often permitted businesses and residences to receive a 5–50 per cent tax credit for purchases of eligible renewable resources (Beck and Martinot, 2004). In the US, the ITC currently covers up to 30 per cent of the cost of a commercial solar or wind project and 10 per cent of the cost of a geothermal project.

ITCs have the benefit of facilitating investment in a specific technology or a suite of technologies. They shift the burden of commercialization to companies and investors instead of concentrating R&D risks with the government. And ITCs, since they operate according to clear percentages, offer investors a guaranteed and predictable source of tax relief.

However, these benefits are more than offset by past failures and shortcomings. ITCs give money for investment in a technology, not its performance, meaning that they often spur research on poorly designed systems. This draws companies into the renewable energy industry that tend to be less experienced and know more about tax management than engineering. In the past, many of these firms have wasted government money by trying to do their R&D 'in the field' to take advantage of the credits. There is a history of ITC-funded wind turbines being sited carelessly and densely, leading to more avian deaths and underperformance, and resulting in lines of motionless machines standing as a testament to hasty development (Grubb, 1990, p530).

In India, the ITC spurred the largest wind power industry among all developing nations but because firms received large economic gains from installing but not operating the wind farms, performance has been poor. Capacity factors have been lower than wind facilities elsewhere, and many wind turbines were reportedly not operational with no efforts made by their developers to repair them (Martinot et al, 2002).

During the 'California Wind Rush' of 1981 to 1985, more than 95 per cent of the world's new wind turbines were installed in California because the ITC strongly incentivized investment but not performance. There, a national ITC covered 25 per cent of wind capital costs and a state ITC covered an additional 25 per cent. With half the cost of each project already covered, developers focused on building wind farms to take advantage of tax credits rather than on building them to produce electricity. When the policy abruptly changed in 1985, manufacturers were unable to improve designs and decrease costs fast enough to justify investment, so a majority of them left the market entirely (Cory et al, 1999).

There are even some cases where the ITC has encouraged scam artists. To capture the ITC rebates, some companies in the US installed non-functioning wind turbines made of cardboard and took pictures of them at the end of the year. These dodgy businesspeople hoped the photos would prevent the Internal Revenue Service (IRS, the national tax collection agency in the US) from disallowing the tax write-offs in case of an audit. If the projects were actually audited, a rare occurrence given the backlog of the IRS at the time, developers then claimed the turbines had been photographed but merely removed for repair. Because the chance of being caught gaming the ITC system was so low, investments in wind energy became a sort of lottery with developers hoping the benefits of getting credit for fake systems outweighed the chances of getting caught (Lotker, 1983). Millions of dollars intended to spur renewable energy could have ended up supporting motionless pieces of cardboard.

In addition, ITCs are expensive and contribute to the volatility of renewable energy markets. In the US, the government spent roughly $6.7 billion on tax credits for energy technologies in 2006 (Metcalf, 2006). Such large expenditures, ironically, can actually *inflate* prices for projects and capital equipment (Kahn, 1996).

Furthermore, the ITC has tended to favour commercial installations. From the start of the credit until 31 December 2008, the ITC in the US capped residential investments in solar energy at $2000 but had no upper limit for commercial installations, creating an asymmetry that heavily favoured centralized and large-scale projects (National Renewable Energy Laboratory, 2009).

Finally, many homeowners and manufacturers around the world lack sufficient income to use the ITC efficiently, since they must have all of the capital up front for investment and can only claim the credit when filing for taxes (Bolinger, 2009). Perhaps because of these reasons, ITCs have played a supplemental, but by no means primary or driving role in investment in renewable energy (Lewis and Wiser, 2005).

Production tax credits (PTCs), unlike ITCs, provide the investor or owner of a qualifying property with an annual tax credit based on the amount of electricity generated by the facility during the course of a year. In the US, this credit has been available to eligible wind, hydropower, landfill gas, municipal solid waste, and biomass facilities (Beck and Martinot, 2004; Bolinger et al, 2009).

The clear difference between the ITC and the PTC is that the latter actually rewards performance and production of energy (much like a FIT). In the US the PTC has provided more efficient labour and greater investment in supply-chain capital associated with wind and solar technologies, including lower risk premiums for manufacturing investment. The PTC has enhanced private R&D expenditures, promoted cost savings by de-linking the US market from the European market (and the unfavourable exchange rate between the Euro and the US dollar), encouraged transportation savings from increased domestic manufacturing of components, and reduced financing charges and fees. Researchers at the Lawrence Berkeley National Laboratory have calculated the benefits of the PTC over ten years at a cost savings of 22 per cent for wind turbines, or a reduction of $380 per installed kilowatt (Wiser et al, 2007a).

The PTC also socializes or distributes the costs of renewable energy projects across all taxpayers. The federal government invests tax revenue in exchange for the benefits of renewable energy, which, in theory, adhere to all citizens (Birgisson and Peterson, 2006). PTC-induced investment has been significant in the US. A survey of renewable energy project developers, capital providers, lawyers, investment bankers, financiers and utility executives found that the PTC accounted for up to 50 per cent of total cash flow in the first ten years of many renewable energy projects and up to one-third of a project's total value over its lifetime (Baratoff et al, 2007). Researchers at Lawrence Berkeley National Laboratory calculated that the PTC was instrumental in attracting nearly $4 billion in wind energy investments in the US in 2006; since the PTC began in 1994, it has helped to incentivize an estimated $13 billion of investment in the renewable energy sector (Wiser et al,

2007a). As perhaps counter-intuitive evidence of the importance of the PTC, wind capacity additions and installations have lagged significantly in years when the PTC faced expiration.

The PTC, however, suffers from some serious disadvantages. The PTC has had the greatest impact in states and localities that have already had incentives such as FITs, RPS and SBCs, implying that it is not enough, by itself, to catalyse investment. Since the PTC can only be used against passive income, it does little to help interested parties (such as farmers, ranchers or urban homeowners) find the funds to invest in wind and solar. These actors lack sufficient capital, and orders for equipment can be cost prohibitive. Most manufacturers of wind turbines, for example, require a 25 per cent deposit upon placement of an order 18–20 months in advance. With each modern turbine at the commercial scale costing $1.6–1.8 million, the PTC does nothing to raise money before a project starts, when it may be most needed (Buis, 2008). It is therefore unclear whether the PTC has been effective in promoting renewable energy that would not have been developed anyway.

PTCs also cost governments millions of dollars in foregone revenue each year, with about $300 million disbursed to projects in the US in 2008. Ninety per cent of these expenditures were for one technology, wind, implying that the PTC does not promote diversification of the renewable resource base. Because the PTC requires direct support from governments, growing budget deficits raise doubts about the long-term commitment of legislators to offer them (Birgisson and Peterson, 2006).

Furthermore, the PTC has distributional income effects, as it results in less government revenue available for other programmes and slightly higher tax rates, creating a greater burden on those in poverty. Some economists and lawyers have argued that the PTC (and other mechanisms like it) has a regressive effect, impacting the financial situation more severely for poor and middle income families from lost government revenue that could be helping them (Ekardt and von Hovel, 2009).

Moreover, the inconsistency of the PTC, mentioned above, has mitigated much of its potential. Private-sector investment has been able to respond to market-based incentives created by the PTC only to the extent that it is credible, lasting and reasonably stable (Mowery, 2006). When Congress failed to restore the credits before the end of 2001, investment in wind turbine projects dropped precipitously. Developers installed only 410MW of new wind turbines in 2002, down from about 1600MW in 2001, and in 2003 it returned to 1600MW (*Megawatt Daily*, 2004). Conflicting policies for wind turbines in the US have created boom and bust cycles within the industry, making it virtually impossible to obtain financing for projects. During this last round of uncertainty, from 2007 to 2008, more than 32,000 direct jobs in manufacturing, construction and operation of wind energy were threatened with layoffs (Lacey, 2008).

Indeed, the same study from Lawrence Berkeley National Laboratory noting that the PTC has reduced costs also noted that it was not nearly as effective as it should have been, given the expenditures involved. The expiration of the PTC and resulting uncertainty slowed wind development in the US, increased reliance on greater foreign manufacturing, and convinced some wind developers to shift to overseas markets. Most developers needed to create complicated finance structures to partner with commercial providers to raise the capital needed to take advantage of the PTC, and industry planners have noted that wind installations in the US are 15–25 per cent more expensive than if the PTC had not been allowed to expire multiple times (Baratoff et al, 2007).

Lastly, the PTC has an inherent bias towards wealthy investors and large corporations. Dr David Toke calls the PTC 'a rich man's FIT' because it must work through actors, such as integrated energy companies or corporations, with high tax liabilities that are able to gain tax benefits from large-scale investments (Toke, 2005). The PTC, simply put, benefits those who already have income. Those who can afford to invest in renewable energy consequently face less financial risk than those that cannot, when incentives should be flowing in the opposite direction (to help those who cannot afford to invest to do so) if distributive justice is a concern (Ekardt and von Hovel, 2009).

Consequently, the PTC functionally excludes individuals and small businesses from developing renewable energy, as it is much more difficult for these groups to acquire capital, and the credits are subject to passive loss rules that limit the types of taxpayers who can use the credit to offset income from other investments (Owens, 2004). As a result, the PTC has promoted the monopolization of the renewables market (especially the wind market) by US firms that have the tax appetite required to utilize the credit, to the detriment of smaller wind developers who become locked out of the market (Owens, 2004). One recent study confirmed that the PTC, while benefiting big energy conglomerates, has actually discouraged community-based projects and investments in small-scale residential systems (Buis, 2008).

9.8 Tendering

In a tendering system (also called a bidding system), renewable energy investors, developers and project owners are invited to apply to bid for a renewable energy contract. The bid that can meet the requirement at the lowest price is awarded the contract, ostensibly guaranteeing the purchase of renewable energy. Renewable electricity is actually sold at market prices, while the difference between the sale and purchase price is financed through a non-discriminatory levy on all domestic electricity consumption (Voogt et al, 2000; Espey, 2001; Meyer, 2003).

The tender system was first developed under the Non-Fossil Fuel Obligation (NFFO) in the UK in 1991, where calls for tenders in relation to energy supply from renewables are made at intermittent intervals. Each renewable technology is given a quota, and the provider with the lowest asking price is given the contract by the UK's 12 public electricity supply companies. A similar competitive bidding process was started in France in 1996, where French regulators defined a reserved market for a given amount of renewable energy, and then organized a competition between renewable providers to allocate this amount.

The tendering system has some advantages. Unlike FITs or tax credits, and akin to an RPS, the government can directly control the amount of renewable energy generation. By allocating contracts on the basis of competitive bidding, providers have an incentive to cut costs to make their bids more attractive. The tendering system also passes on such savings to tax payers and consumers, rather than concentrating profits in corporations or investors.

However, the tendering system, because of its considerably lower purchase prices, also has a host of sobering problems:

1. It has tended to hurt providers by persuading them to reduce margins to make their bid attractive. Such lower margins can damage their investment capabilities and in extreme situations lead to bankruptcy. In the UK, the tendering system has been accused of not promoting local renewable energy development, as a majority of providers have imported technology to meet the standard from out of the country. One study looking at the various policy options to promote renewables noted that tendering systems did not usually provide long-term market stability or profitability, and also resulted in long lead times between lending and project completion, and fierce competition among project developers to the point that it was counterproductive (Lewis and Wiser, 2005). Another study found that only 30 per cent, *less than one-third*, of contracted capacity under the NFFO was actually installed. By 2003, only 960MW had been installed out of 3270MW awarded. The explanation is that bidders went too low in order to win their contracts, and then backed away from installing projects to avoid going bankrupt (Butler and Neuhoff, 2005). The same trend has occurred in other tendering systems. France's programme saw just 70 MW built out of 300MW contracted and of that, just 30MW were operating in 2005 (Butler and Neuhoff, 2005).
2. Tendering does not usually create sustainable renewable energy growth. Because it is based on auctions and calls for bids, tendering systems tend to cause stop-and-go development cycles, much like the PTC in the US (Gipe, 2006). The erratic announcement of tenders means that firms often cannot know when the next round would be announced, complicating investment decisions.
3. The intense price competition resulting from tendering favoured large incumbent renewable energy developers and suppliers, usually state-owned enterprises without profit motives, at the expense of independent providers and small firms. One report concluded that tendering does not stimulate development of a diverse renewable energy market and can actually worsen anti-competitive

practices in electricity markets (Government of South Australia, 2007). Such challenges with tendering often translate into higher risks and therefore higher overall costs. The bulk of financing for many tendering (and quota) systems has been provided by large banks and insurance companies that must partner with project developers through specialized financing structures needed to mitigate uncertainty and risk (Bolinger et al, 2009).

4   Since the chance of winning bids in a competitive market is rather low, many investors and potential power providers simply decide not to participate (Madlener and Stagl, 2005). This limited number of bidders allows the tendering system to be easily gamed. For example, in some tendering schemes a small number of large players have 'gamed' bid prices (but never intended to actually complete projects) to block out competitors. Similarly, state-owned enterprises can commit to unreasonably low prices to win contracts since their lack of profitability means that they can force ratepayers to subsidize the eventual projects (Madlener and Stagl, 2005).

9.9 THE SUPERIORITY OF FITs

Of course, this entire book is about why FITs are superior to other mechanisms, but to those wanting comparative analysis, the benefits of FITs have been confirmed by hundreds of independent studies and empirical findings.

To select just a few recent examples, the International Energy Agency recently examined the effectiveness of renewable energy policies in 35 countries including members of the EU and US along with Brazil, Russia, India, China and South Africa (IEA, 2008). Together these countries account for 80 per cent of total commercial renewable power generation around the world, 77 per cent of renewable heating and cooling, and 98 per cent of renewable transport fuel production. The study noted that the group of four countries with highest effectiveness – Germany, Spain, Denmark and Portugal – *all* used FITs to encourage wind and solar development. The study praised FITs for promoting high rates of investment stability, their simple framework with low administrative barriers and costs, and their tendency to have favourable grid access conditions. The study concluded that the average remuneration rates for FIT programmes (US$0.09–0.11/kWh) were far more cost-effective than any other mechanism (quota systems and tradable certificates averaged US$0.13–0.17/kWh).

Another study compared FITs with other mechanisms such as quotas and certificates (Fouquet and Johansson, 2008). The assessment noted that quota systems such as RPS and the Renewables Obligation end up restricting market size to the extent of the quota, in essence creating a ceiling on renewable energy penetration. FITs, by contrast, have no such cap and therefore no such restriction, and cause a faster shift towards distributed resources and smaller-scale systems installed by smaller firms, which improves parity and access for industry development. The study found that REC schemes are intrinsically volatile since their value changes over time. The oversupply of RECs can cause their price to drop precipitously, leading shrewd investors to refrain from decisions and purchases that would affect certificate prices too much, mitigating investment opportunity. The study concluded that compared with other electricity mechanisms for renewables in all 27 EU member states, FITs were by far the most effective, delivering larger and faster penetration at lower cost than certificates and quotas, mitigating investor risks, and promoting innovation in less mature technologies.

Yet another study compared the effectiveness of different support schemes and the level of financial support they required for wind energy in the EU (Ragwitz et al, 2006). The study noted that the level of financial support for countries relying on quota obligations and RECs was much higher than in those with only FITs. In three countries that rely primarily on quota systems, Belgium, Italy and the UK, the study discovered higher cost for the programme but lower growth rates. Countries with FITs, by contrast, had higher growth rates but lower costs. The study concluded that FITs are more effective than other policy mechanisms for four reasons:

1 They drive down capital costs, which are much lower (and achieve quicker price reductions) in countries with FITs than in countries with quota obligations.
2 They promote a diversified portfolio of technologies and industrial sectors.
3 They minimize electricity costs in two ways: guaranteed tariffs lower risk and therefore need less capital acquired at lower interest rates; and programmes with degression (also called stepped tariffs) reduce the potential for surplus profits and encourage efficiency and lower manufacturing costs over time.
4 They encourage market competition as they do not suffer many of the liquidity and market abuse problems of the REC system, and encourage competition among manufacturers rather than investors, leading to better market conditions for building and deploying renewables.

A fourth assessment investigated the performance of FITs in Germany and Spain versus the Renewables Obligation in the UK. The assessment found that the German market was the least risky for renewable energy development as it isolated suppliers from market prices and risks. Quota and tradable credit schemes, by contrast, exacerbated the risk of electricity price increases (Klessmann et al, 2008).

Most telling is the perspective of actual investors, economists and financial firms. Ernst & Young, a global financial conglomerate, recently argued that they believed FITs are more cost-effective than other policy mechanisms, and would prefer to see FITs in places their clients want to invest. They ranked the German market the 'most attractive' for renewable energy investment in the world, ahead of the US, because of its FIT. They also projected that in France the FIT has been responsible for 10.5TWh of renewable power in 2006 at a cost of €124 million (displacing CO_2 at a cost of €39.50 per tonne) while the Renewables Obligation in the UK brought roughly the same amount of power, 12.9TWh, but cost £611 million (or a cost of €86 per tonne of CO_2), more than *twice* as much (Chadha, 2009). The comparison between Germany and the UK is just as favourable for FITs. In 2007, Germany generated 72.7TWh of renewable power at cost of 0.045/kWh (average) with their FIT, while the UK generated 18.1TWh at a cost of 0.056/kWh (average) with their Renewables Obligation. The famous *Stern Review* on the financial costs of global climate change, written by the former chief economist at the World Bank Nicholas Stern, also noted that FITs achieved 'larger deployment [of renewable energy systems] at lower costs' compared to tradable certificates and quota schemes (Stern, 2006).

The list could go on and on, but the authors trust that the point has been made. Besides these economic advantages, FITs also encourage the investment of a large number of potential producers, ranging from homeowners and small to medium-sized companies to large utilities. This way, competition in the power production sector can be increased, making it more competitive and democratic at once.

9.10 Conclusion

Perhaps readers at this point are starting to feel a bit like Meno in Plato's dialogue. If you want to promote renewable energy, you can count on the goodwill of people to voluntarily purchase it through green power programmes. You can enable homeowners and business leaders who want to sell power back to the grid through net metering programmes. You can create complicated and often convoluted markets for tradable certificates. You can fund basic research and development in a particular technology and hope it pays off in the long run. You can force electric utilities to produce a certain quantity of renewable energy by a given date but leave the price up to the market. You can create a miniscule tax on every kWh of electricity generated and use some of the funds to invest in renewable energy. You can give large companies and savvy homeowners tax credits to partially offset their expenditures on renewable energy equipment or performance. You can create a complex bidding and auction system for renewable energy contracts. Or you can merely pay for renewable energy through a FIT.

Almost every one of these alternative mechanisms tries to create price incentives for renewable electricity, some more effectively than others, some more directly than others. Many have transaction costs that serve to benefit particular firms, consultants and stakeholders, often at the expense of other firms and to the exclusion of small-scale distributed systems and by inflating electricity prices. Most do not promote dynamic efficiency and instead focus on the cheapest and most mature technologies.

Only FITs, however, have minimal transaction costs, promote diversification and lower electricity prices all at once. And only FITs are backed by scores of

OTHER SUPPORT SCHEMES 179

independent studies comparing many of the mechanisms discussed in this chapter against each other. The basic lesson appears to be simple: policy makers need no longer remain in a perpetual state of *aporia*. There is one option superior to all others, and it is a FIT.

REFERENCES

Baratoff, M. C., Black, I., Burgess, B., Felt, J. E., Garratt, M. and Guenther, C. (2007) *Renewable Power, Policy, and the Cost of Capital: Improving Capital Market Efficiency to Support Renewable Power Generation Projects*, United Nations Environment Programme/BASE Sustainable Energy Finance Initiative, Paris

Beck, F. and Martinot, E. (2004) 'Renewable energy policies and barriers', in C. Cleveland (ed.) *Encyclopedia of Energy*, Elsevier Science, San Diego, vol 5, pp365–383

Berendt, C. (2006). 'A state-based approach to building a liquid national market for renewable energy certificates: The REC-EX model', *The Electricity Journal*, vol 19, no 5, pp 54–68

Bergey Windpower (1999) *Net Metering and Related Utility Issues*, www.bergey.com/School/FAQ.Net-Metering.html, pp1–2

Bird, L. A., Cory, K. S. and Swezey, B. G. (2008) *Renewable Energy Price-Stability Benefits in Utility Green Power Programs*, NREL/TP-670-43532, National Renewable Energy Laboratory, Golden, CO

Bird, L., Hurlbut, D., Donohoo, P., Cory, K. and Kreycik, C. (2009) *An Examination of the Regional Supply and Demand Balance for Renewable Electricity in the United States through 2015*, NREL/TP-6A2-45041, National Renewable Energy Laboratory, Golden, CO

Birgisson, G. and Peterson, E. (2006) 'Renewable energy development incentives: Strengths, weaknesses and the interplay', *The Electricity Journal*, vol 19, no 3, pp40–51

Bolinger, M. (2009) *Financing Non-residential Photovoltaic Projects: Options and Implications*, LBNL-1410E, Lawrence Berkeley National Laboratory, Berkeley, CA

Bolinger, M., Wiser, R. and Fitzgerald, G. (2005) 'An overview of investments by state renewable energy funds in large-scale renewable generation projects', *The Electricity Journal*, vol 18, no 1, pp78–84

Bolinger, M., Wiser, R., Cory, K. and James, T. (2009) *PTC, ITC, or Cash Grant? An Analysis of the Choice Facing Renewable Power Projects in the United States*, NREL/TP-6A2-45359, National Renewable Energy Laboratory, Golden, CO

Bradbrook, A. J., Lyster, R. and Ottinger, R. L. (2005) *The Law of Energy for Sustainable Development*, Cambridge University Press, Cambridge, p113

Buis, T. (2008) *Concerning the Renewable Energy Economy: A New Path to Investment, Jobs, and Growth*, Testimony before the House Select Committee on Energy Independence and Global Warming, US Government Printing Office, Washington, DC

Butler, L. and Neuhoff, K. (2005) *Comparison of FIT, Quota, and Auction Mechanisms to Support Wind Power Development*, CMI Working Paper 70, www.electricitypolicy.org.uk/pubs/wp/ep70.pdf

California Public Utilities Commission (2007) *Progress on the California Renewable Portfolio Standard as Required by the Supplemental Report of the 2006 Budget Act*, CPUC Report to the Legislator, Sacramento, CA

Carbon Trust (2007) *Policy Frameworks for Renewables: Analysis on Policy Frameworks to Drive Future Investment in Near and Long-Term Renewable Power in the UK*, The Carbon Trust and L. E. K. Consulting, London

Center for Resource Solutions (2001) *Interaction Between RPS and SBC Policies*, Center for Resource Solutions, Washington, DC, p2

Chadha, M. (2009) 'Germany overtakes us as most attractive market for renewable energy', *Energy Investment*, 4 January, http://redgreenandblue.org/2009/01/04/germany-overtakes-us-as-the-most-attractive-market-for-renewable-energy-investment/

Coenraads, R., Reece, G., Voogt, M., Ragwitz, M., Resch, G., Faber, T., Haas, R., Konstantinaviciute, I., Krivosik, J. and Chadim, T. (2008) *Progress: Promotion and Growth of Renewable Energy Sources and Systems*, Ecofys, Fraunhofer Institute, Energy Economics Group, LEI and SEVEn; Utrecht

Cory, K. S., Bernow, S., Dougherty, W., Kartha, S. and Williams, E. (1999) *Analysis of Wind Turbine Cost Reductions: The Role of Research and Development and Cumulative Production*, Presentation to the American Wind Energy Association WINDPOWER Conference

Coughlin, J. and Cory, K. (2009) *Solar Photovoltaic Financing: Residential Sector Deployment*, NREL/TP-6A2-44853, National Renewable Energy Laboratory, Golden, CO

Dinica, V. (2006) 'Support systems for the diffusion of renewable energy technologies – an investors perspective', *Energy Policy*, vol 34, pp461–480

Duke, R., Williams, R. and Payne, A. (2005) 'Accelerating residential PV expansion: Demand analysis for competitive electricity markets', *Energy Policy*, vol 33, p1916

Ekardt, F. and von Hovel, A. (2009) 'Distributive justice, competitiveness, and transnational climate protection: "One human – one emission right"', *Carbon & Climate Law Review*, vol 3, no 1, pp102–113

Espey, S. (2001) 'Renewables portfolio standard: A means for trade with electricity from renewable energy sources?', *Energy Policy*, vol 29, pp557–566

FERC (1995) *Promoting Wholesale Competition through Open Access Non-discriminatory Transmission Services by Public Utilities and Recovery of Stranded Costs by Public Utilities and Transmitting Utilities, Notice of Proposed Rulemaking and Supplemental Notice of Proposed Rulemaking*, Docket Nos. RM95-8-000 and RM94-7-001, Washington, DC 1995

Fouquet, D. and Johansson, T. B. (2008) 'European renewable energy policy at a crossroads – focus on electricity support mechanisms', *Energy Policy*, vol 36, no 11, pp4079–4092

Gallagher, K. S., Frosch, R. and Holdren, J. P. (2004) *Management of Energy Technology Innovation Activities at the US Department of Energy*, Report to the Belfer Center for Science and International Affairs, Cambridge, MA

Gipe, P. (2006) *Renewable Energy Policy Mechanisms*, Wind Works Organization, Tehachapi, CA

Gomez, A. (2009) 'Going green can eat a lot of green,' *USA Today*, 5 May, p3A

Government of South Australia (2007) *South Australia's Feed-In Mechanism for Residential Small-Scale Solar Photovoltaic Installations*, Government of South Australia, Adelaide, South Australia

Grubb, M. J. (1990) 'The cinderella options: A study of modernized renewable energy technologies part 1 – A technical assessment', *Energy Policy*, vol 18, no 6, pp525–542

Haddad, B. and Jefferiss, P. (1999) 'Forging consensus on national renewables policy: the renewables portfolio standard and the National Public Benefits Trust Fund', *The Electricity Journal*, vol 12, no 2, p70

Haynes, R. (2007) 'Not your mother's net metering: recent policy evolution in U.S. States', Presentation to the Solar Power 2007 Conference in Long Beach, CA, available at www.dsireusa.org/documents/PolicyPublications/Solar%20Power%202007.ppt

IEA (2008) *Deploying Renewables: Principles for Effective Policies*, OECD/IEA, Paris

IEA (2009) *Beyond 2020 World Data Services*, International Energy Agency, available at http://wds.iea.org/WDS/Common/Login/login.aspx

Interstate Renewable Energy Council (2008) *Public Benefits Funds for Renewables*, www.dsireusa.org/incentives/index.cfm?EE=/&RE=1, p1

Jaccard, M. (2004) 'Renewable portfolio standard', in C. Cleveland (ed.) *Encyclopedia of Energy*, Elsevier Science, San Diego, vol 5, pp413–421

Jacobsson, S., Bergek, A., Finon, D., Lauber, V., Mitchell, C., Toke, D. and Verbruggen, A. (2009) 'EU renewable energy support policy: Faith or facts?', *Energy Policy*, vol 37, pp2143–2146

Jørgensen, B. H. (2006) *Energy R&D Contributions to Future Economic Growth and Social Welfare*, Seminar on Nordic Research and Innovation Cooperation with Estonia, p5

Kahn, E. (1996) 'The Production Tax Credit for wind turbine powerplants is an ineffective incentive', *Energy Policy*, vol 24, no 5, pp427–435

Klessmann, C., Nabe, C. and Burges, K. (2008) 'Pros and cons of exposing renewables to electricity market risks – a comparison of the market integration approaches in Germany, Spain, and the UK', *Energy Policy*, vol 36, pp3646–3661

Lacey, S. (2008) 'Building a FIT renewable energy market in the U.S.', *Renewable Energy World*, 10 March www.renewableenergyworld.com/rea/news/article/2008/03/building-a-fit-renewable-energy-market-in-the-u-s-51798

Langeraar, J.-W. and de Vos, R. (2003) 'Guarantee of origin: The proof of the pudding is in the eating', *Refocus*, July/August, pp62–63

Lauber, V. (2004) 'REFIT and RPS: Options for a harmonized community framework', *Energy Policy*, vol 32, pp1405–1414

Lewis, J. and Wiser, R. (2005) *Fostering a Renewable Energy Technology Industry: An International Comparison of Wind Industry Policy Support Mechanisms*, LBNL-59116, Lawrence Berkeley National Laboratory, Berkeley, CA

Lotker, J. (1983) *Wind Energy Commercialization: A Premature Retrospective*, Presentation to the Wind Workshop VI, Minneapolis, MN

Madlener, R. and Stagl, S. (2005) 'Sustainability-guided promotion of renewable electricity generation', *Ecological Economics*, vol 53, pp147–167

Martinot, E., Chaurey, A., Lew, D., Moreira, J. E. and Wamukonya, N. (2002) 'Renewable energy markets in developing countries', *Annual Review of Energy and the Environment*, vol 27, pp309–348

Megawatt Daily (2004) 'Wind group says loss of tax credits stalls 1,000 MW', *Megawatt Daily*, no 9, 6 January, p8

Menz, F. C. and Vachon, S. (2006) 'The effectiveness of different policy regimes for promoting wind power: Experiences from the States', *Energy Policy*, vol 34, pp1786–1796

Metcalf, G. E. (2006) *Federal Tax Policy towards Energy*, National Bureau of Economic Research, Cambridge, MA

Meyer, N. I. (2003) 'European schemes for promoting renewables in liberalized markets', *Energy Policy*, vol 31, pp665–676

Mowery, D. C. (2006) *Lessons from the History of Federal R&D policy for an Energy ARPA*, Statement before the Committee on Science, US House of Representatives, US Government Printing Office, Washington, DC, p4

National Renewable Energy Laboratory (2009) *Solar Leasing for Residential Photovoltaic Systems*, NREL/FS-6A2-43572, Golden, CO

Network for New Energy Choices (2007) *Freeing the Grid 2007*, Network for New Energy Choices, New York, NY

Ontario Ministry of Energy (2008) *Net Metering in Ontario, 2008*, Ontario Ministry of Energy, Ontario, Canada, pp1–2

Owens, B. (2004) 'Beyond the Production Tax Credit (PTC): Renewable energy support in the US', *Refocus*, vol 5, no 6, November/December, pp32–34

Rader, N. and Hempling, S. (2001) *The Renewables Portfolio Standard: A Practical Guide*, National Association of Regulatory Utility Commissioners, Washington, DC

Ragwitz, M., Held, A., Resch, G., Faber, T., Huber, C. and Haas, R. (2006) *Monitoring and Evaluation of Policy Instruments to Support Renewable Electricity in EU Member States*, Fraunhofer Institute, Frieberg

Ragwitz, M., del Río Gonzalez, P. and Resch, G. (2009) 'Assessing the advantages and drawbacks of government trading of guarantees of origin for renewable electricity in Europe', *Energy Policy*, vol 37, pp300–307

Refocus (2003) 'Guarantee of origin: A major breakthrough for the internal green energy market', *Refocus*, March/April, pp60–61

REN21 (2008) *Renewable Energy Policy Network for the 21st Century, Renewables 2007*, Global Status Report, Washington, DC, p7

REN21 (2009) *Renewables Global Status Report: 2009 Update*, REN21 Secretariat, Paris, www.ren21.net/pdf/RE_GSR_2009_update.pdf

Rickerson, W. and Grace, R. C. (2007) *The Debate over Fixed Price Incentives for Renewable Electricity in Europe and the United States: Fallout and Future Directions*, Heinrich Boll Foundation, Washington, DC

Sovacool, B. K. (2008a) 'A matter of stability and equity: The case for federal action on renewable portfolio standards in the U.S.', *Energy and Environment*, vol 19, no 2, pp241–261

Sovacool, B. K. (2008b) 'The best of both worlds: Environmental federalism and the need for federal action on renewable energy and climate change', *Stanford Environmental Law Journal*, vol 27, no 2, pp397–476

Sovacool, B. K. (2009) 'The importance of comprehensiveness in renewable electricity and energy efficiency policy', *Energy Policy*, vol 37, pp1529–1541

Sovacool, B. K. and Cooper, C. (2007) 'Big is beautiful: The case for federal leadership on a national renewable portfolio standard', *The Electricity Journal*, vol 20, no 4, pp48–61

Sovacool, B. K. and Cooper, C. (2007–2008) 'The hidden costs of state Renewable Portfolio Standards (RPS)', *Buffalo Environmental Law Journal*, vol 15, no 1–2, pp1–41

Sovacool, B. K. and Cooper, C. (2008) 'Congress got it wrong: The case for a national Renewable Portfolio Standard (RPS) and implications for policy', *Environmental and Energy Law and Policy Journal*, vol 3, no 1, pp85–148

Stern, N. (2006) *Stern Review: The Economics of Climate Change*, Cambridge University Press, Cambridge, Part IV: Policy responses for mitigation

Toke, D. (2005) 'Are green electricity certificates the way forward for renewable energy? An evaluation of the United Kingdom's Renewables Obligation in the context of international comparisons', *Environment and Planning C: Government and Policy*, vol 23, pp361–374

Tonn, B., Healy, K. C., Gibson, A., Ashish, A., Cody, P., Beres, D., Lulla, S., Mazur, J. and Ritter, A. J. (2009) 'Power from perspective: Potential future United States energy portfolios', *Energy Policy*, vol 37, pp1432–1443

US Department of Energy (2006) *Small Wind Electric Systems: A Consumer's Guide*, DOE, Washington, DC, p18

US Department of Energy (2008) *Annual Report on US Wind Power Installation, Cost, and Performance Trends: 2007*, US Department of Energy, Washington, DC

US Energy Information Administration (2007) *Annual Energy Outlook*, US Department of Energy, Washington, DC, pp3–10

US Government Accountability Office (2005) *National Energy Policy: Inventory of Major Federal Energy Programs and Status of Policy Recommendations*, GAO-05-379, United States GAO Report to Congress, pp1–5

US Government Accountability Office, Department of Energy (2006) *Key Challenges Remain for Developing and Deploying Advanced Energy Technologies to Meet Future Needs*, GAO-07-106, GAO, Washington, DC, p5

US House of Representatives Committee on Science (2006) *Renewable Energy Technologies – Research Directions, Investment Opportunities, and Challenges to Commercial Applications in the United States and Developing World*, Hearing Charter, 9pp

Voogt, M., Boots, M. G., Schaeffer, G. J. and Martens, J.W. (2000) 'Renewable electricity in a liberalized market–the concept of green certificates', *Energy and Environment*, vol 11, no 1, pp65–80

Washington Utilities and Transportation Commission (1994) *DSM Tariffs UE-941375 and UE-941377*, Washington Utilities and Transportation Commission, Olympia, WA

Wiser, R. H. (2007) '*State Policy Update: A Review of Effective Wind Power Incentives*', Presentation to the Midwestern Wind Policy Institute, Ann Arbor, MI

Wiser, R. and Barbose, G. (2008) *Renewables Portfolio Standards in the United States: A Status Report with Data through 2007*, LBNL-154E, Lawrence Berkeley National Laboratory, Berkeley, CA

Wiser, R., Bolinger, M., Milford, L., Porter, K. and Clark, R. (2002) *Innovation, Renewable Energy, and State Investment: Case Studies of Leading Clean Energy Funds*, LBNL-51493, Lawrence Berkeley National Laboratory, Berkeley, CA

Wiser, R., Bolinger, M. and Barbose, G. (2007a) 'Using the federal production tax credit to build a durable market for wind power in the United States', *The Electricity Journal*, vol 20, no 9, pp77–88

Wiser, R., Namovicz, C., Gielecki, M. and Smith, R. (2007b) 'The experience with renewable portfolio standards in the United States', *The Electricity Journal*, vol 20, no 4, May, pp8–20

10
Campaigning for FITs

This chapter rounds out the book by looking at how one may go about campaigning for feed-in tariffs (FITs). One might reasonably ask why campaigns are needed if the advantages of FITs seem so clear. Well, as alluded to in previous sections, the energy and electricity industry is highly politicized, with conventional energy companies investing billions of dollars into fossil-fuelled and nuclear systems every year and spending as much as US$255 million per year making 'contributions' to political parties and campaigns in the US alone (Sovacool, 2008, p230). Amory Lovins, the energy efficiency guru and committed advocate of pursuing a 'soft path' to energy policy by promoting renewables, once remarked that 'hell will freeze over first' before all politicians agree about energy issues (Sovacool, 2008, p235). In a world where oil companies spend millions of dollars purposely funding false anti-climate science, nuclear companies create clever organizations and media campaigns to manipulate public opinion, and the coal industry gives us the 'Clean Coal Carolers' at Christmas time to sing *Deck the Halls (With Clean Coal)*, FITs – and renewable energy – will have their opponents. This chapter clarifies some of this opposition, and how it can be fought.

10.1 WHAT ARE WE UP AGAINST?

When working in areas as lucrative and indispensable as energy supply and use, it does not take long to realize that there are many stakeholders driven by self-interest to prioritize profits over public service. The blockers of renewable energy (and of FITs by extension) are by and large those who would lose out financially from their deployment and implementation. Keeping with our holiday analogies, the phrase to keep in mind when dealing with this subject is 'Turkeys don't vote for Christmas' (or Thanksgiving, for our North American readers). There are a lot of energy turkeys out there, they run the show at present in most places, and they will vigorously oppose any talk of Christmas, or any such festivities which they do not organize and profit from exclusively. (The analogy to Christmas is apt as FITs do bring many gifts, while the energy turkeys literally bring lumps of coal.) Energy turkeys will state a great many reasons to oppose FITs, but these become transparently specious when one realizes that they are simply trying to protect their interests, not ours.

If we accept this view (and there are few who do not, turkeys aside), the question then becomes one of strategy. How might these powerful, influential interests be overcome for the greater good? The same patterns of opposition have been witnessed time and again, and successful methods of overcoming them have emerged. Let us look at some examples of opposition.

The debate over SRECs (Solar Renewable Energy Certificates) in the US is a clear example of where political pressures have created government policy that favours large interests at the expense of smaller players (Lacey, 2008). Such interests were able to successfully utilize their political influence to bring about policies favourable to themselves, and continue to push hard in this direction.

Recent disasters in the financial and automotive sectors have illustrated what happens when dominant businesses get 'too big to fail'. Ecology shows us that diversity is fundamental to the creation of strong, healthy, resilient systems. The opposite is a dependency trap, into which taxpayers have fallen with the banks and auto manufacturers. Energy companies are no different. We can expect that many of these companies will resist improved self-reliance on the part of citizens and communities to produce their own energy. It means fewer profits for them (Mendonça, 2007, pp17–18). Consider one story from the US, typical of such resistance. Managers from a conventional electricity utility spent seven years trying to stop one man from connecting a small-scale wind turbine to the company's distribution system. The man, a small farmer from rural Iowa, had to hire a lawyer and appeal to the state court system and the Federal Energy Regulatory Commission (FERC). FERC ruled in favour of the farmer and scolded the cooperative's managers for deliberately disconnecting the family, for using delaying tactics, and for arguing disingenuously to the courts (Sovacool, 2008, p144).

In Europe, the German Electricity Association (speaking for the major energy companies) attempted to have the FIT removed on a variety of grounds, including arguments connected to cost, infrastructure issues, market rules and legal statutes. While they were comprehensively rebutted, and failed on all counts, it does not appear that they have given up their quest to have a policy enacted which suits their members better and allows them to dominate the market unhindered (Mendonça, 2007, pp39–42).

10.2 How Do We Overcome these Blockages?

Two things can be done to overcome these problems of vested interests: improve the decision-making process in terms of participation and transparency, and make clear who the beneficiaries of certain policies are. To give just two examples of research and advocacy related to decision making, Frances Moore Lappé has done a great deal of work on reform of democratic processes, using the term 'living democracy'

(Lappé, 2007) and Frede Hvelplund uses the term 'innovative democracy' to describe similar processes of broad stakeholder engagement and collaborative decision making, with particular regard to renewable energy (Mendonça et al, 2009).

The analysis of policies can be especially powerful in revealing who stands to benefit most. The criteria below are a useful start towards thinking systematically about this in terms of renewable energy policies and FITs:

- Local acceptance – how does the policy account for and influence this?
- Equity – how open is it to investment from all sources?
- Simplicity – how simple and intelligible is the policy?
- Benefits – whose interests are furthered or protected?
- Transition – how does the policy link to previous and supporting policies?
- Policy creation – how are NGOs and new technologies represented in policy creation processes?
- Infrastructure policy – to what extent is a new infrastructure for renewable energy technologies established?

These criteria emphasize the importance of broader engagement in decision making, a wider distribution of benefits, promoting decentralized generation, and encouraging diverse ownership structures and the establishment of renewables-friendly infrastructure. The main goals of these criteria could be labelled 'stability', 'participation' and 'transparency'. Interestingly FITs meet all three criteria. They provide the long-term stability necessary for investor confidence, as well as participation through being open to all potential investors. Transparency arises from the policy-making process, as well as from analysis of beneficiaries. Expert analysis should yield very quickly a clear view of who the beneficiaries are, and how widely the benefits are distributed.

One of the key benefits of greater participation is the engagement of citizens and communities concerned about climate change and environmental protection. This comment from Henning Holst, a wind power consultant in Germany, captures a sense of these benefits in the real world:

> *What happened after the EFL [Electricity Feed Law] is that many individuals have invested in local wind power schemes. These people have become 'energy experts', so people are much more aware about wind energy. Now everybody is aware that electricity doesn't just come out of a plug. Because there are thousands of investors in wind energy there is a strong lobby for good conditions for wind energy in the future* (Toke, 2005).

This raises the next point, which is that of the widening of the stakeholder group which provides political support for more green legislation. We can see from the FIT advantages shown in Table 0.1 in the Introduction that the increasing involvement of ordinary people in the energy production process can drive more awareness of energy consumption, and it can also increase the level of support for more green policies. Arguing on behalf of FITs is thus a matter of finding allies who share this understanding of democracy.

The next section offers a guide to setting out a successful campaign for the introduction of a FIT. The steps will not necessarily occur in the order they are presented, or be necessary in their entirety for all countries or communities. They may well reappear in different forms throughout the process, for different audiences and purposes. The nature of a campaign for FITs will be influenced by many factors, including the government's appetite for renewable energy; the power of vested interests; the vision, organizational capacity and influence of the convening group; the size, make-up and determination of the coalition; the agreement on non-negotiable outcomes; the willingness of coalition members to cooperate and to compromise; and funding.

10.2.1 Learning

The first step is to learn about the policy itself. What is it? Why does it work so well? Who has most experience with it? What has it achieved? This book provides an answer to each of these questions, and we need not reiterate them here. As we have seen, there are political, legislative, technical, social, financial, economic, cultural and planning dimensions to FITs, and coalitions must address them all.

As described in various chapters of this book, design is important, but supporting conditions are essential to the effective implementation of the law. These include the condition of grid infrastructure and planning laws. If the latter are a barrier to deployment, they must be taken on in tandem. Further, the development of a FIT is an ideal time to look at the entire energy system, with particular regard to smart grids, smart meters, demand management and electric vehicles – suggesting the potential for the creation of an umbrella bill which accounts for all of these issues. Those campaigning for FITs would benefit greatly from becoming experts about energy issues in their own community – they would be able to understand the application of FITs more clearly. Readers who have read this far are already on a good trajectory.

For access to many and varied studies, publications, guidance and advocates, some excellent web-based materials and potential allies can be found at these sites:

- Alliance for Renewable Energy – www.allianceforrenewableenergy.org
- Alliance for Rural Electrification – www.ruralelec.org
- German Federal Environment Ministry's renewable energy pages – www.erneuerbare-energien.de/inhalt/3860/

- International Feed-in Cooperation – www.feed-in-cooperation.org
- International Renewable Energy Agency (IRENA) – www.irena.org
- Policy Action on Climate Toolkit (PACT) – http://onlinepact.org
- REFIT NZ – http://refit.org.nz
- RES Legal – http://res-legal.de/en/
- Wind Works – http://wind-works.org

10.2.2 Finding allies

No matter who you are, if there is going to be a fight to see FIT policies enacted, advocates will need to look for allies. These may include:

- academics, research groups, think tanks, independent experts;
- environmental non-profits/ NGOs/ charities;
- legislators/parliamentarians;
- manufacturers;
- businesses;
- utilities (often municipal utilities first);
- renewable energy trade associations;
- community/grassroots groups, co-operatives;
- first nations/indigenous groups;
- investors, investment banks;
- chambers of commerce;
- development agencies;
- labour unions;
- house builders' associations;
- landowners' associations;
- landlords' associations;
- retailers' associations
- fuel poverty groups; and
- celebrities.

This list is drawn from a variety of groups and individuals active on the subject in various countries. There is no reason why, for example, the military and the emergency services would not benefit also. The US is already working on powering some of its military bases with renewable energy. While this may seem contrary to the values of many espousing the use of renewables, it must be remembered that many advances in solar arose originally from the US space programme. The green part of the 2009 economic stimulus package, the American Recovery and Reinvestment Act, contained a US$350 million provision for Department of Defense Research into using renewable energy to power weapons systems and military bases. The former director of the CIA, James Woolsey, argues in favour of

FITs from the perspective of facilitating distributed generation, which is far more resilient against disruption from terror attacks.

Unfortunately, you may find that those you assume to be natural allies turn out to be 'turkeys'. Some of the types of organizations on our list may be opposed to FITs because their interests diverge, for the reasons set out above. However, one must not become too discouraged, as by their very nature more people can gain from FITs than can possibly lose from them, and a good coalition will draw from these increasing numbers of people as it progresses.

10.2.3 Workshops, conferences and fact-finding

Holding workshops is practical and useful for a variety of reasons. They help to develop knowledge for the general and local context, attract supporters, identify opponents and bring together individuals in person. A convening organization or group will generally organize the gathering, and there may be a series of them, as they have uses throughout the different stages of the process. Inviting acknowledged experts to make presentations will naturally help to create the requisite interest in the event to make it a success (see the PACT site above for a roster of international experts). This is a self-selecting process in some ways, in that often both supporters and opponents will be present at these events, making it clear who is on which side. It can be very productive to argue out the pros and cons for the jurisdiction in question.

Workshops are generally very positive, and will bring together those who wish to see the policy in place. There will often be a general understanding of the policy and its benefits among those present, and presentations should be geared towards all levels of competency. This helps to give those witnessing the presentations an idea of their level of understanding, and the variety of competencies necessary to take this subject through the full process. Even one day events can be useful in terms of getting organized, launching a campaign, or clinching a campaign by bringing in dignitaries that will attract attention in the media and give personality and authenticity to FITs.

One of the most effective methods of establishing interest is to take groups overseas to see the effects of the legislation for themselves. Fact-finding trips to Germany and Spain have proved particularly effective, and places such as the city of Freiburg in Germany and the regions of Granada and Navarra in Spain have seen many such visits. The opportunity to actually visit real infrastructure and to meet the people making it happen is a powerful thing, and has already aided the international spread of FITs.

10.2.4 Fund-raising and communications

Fund-raising should be a priority for any group setting out on a campaign. There may be the possibility to 'host' a workshop or seminar through a trade association, non-governmental organization (NGO) or other institution, but funds will still be required for other events, publications and communications, or even for specially hired staff to coordinate the campaign. Some manufacturers may commit funds, and there are many foundations that can be convinced of the case for an accelerated switch to renewable, secure, democratized energy. Chapter 1 sets out the enormous potential for lucrative spin-off benefits from this transition.

Publications and communications are essential throughout the process, and can also play a key role in raising the awareness needed for fund-raising. Publications can include general fact sheets on FITs; position papers (setting out the stance of various coalition members); individual, joint or commissioned research on what kind of FIT options would work for the jurisdiction in question; working papers or reports on FITs and local job creation, and so on. Different audiences can be targeted with tailored publications, addressing them at their level, on the issues most pertinent to them.

Many forms of communication can be done cheaply and quickly. Possibilities range from internal organization of the coalition via email list servers and newsgroups, to external press releases, letters to the editor, comment pieces, blog entries, Facebook group messages, Twitter 'tweets' and news items. A coalition website is a very effective tool, some examples of which are listed above. It is particularly handy for following national and regional debate on the subject, such as media articles, public events, legislative news and so on.

Part of the job of communications will be to refute the anti-renewables or anti-FIT propaganda that gets placed in the media, either through news reports or 'studies'. At the time of writing, a report on the negative impacts of renewable energy support on jobs in Spain is being touted as an empirical argument against the same types of support in the US. This report is in the process of being systematically refuted by experts in North America and elsewhere. As one might have suspected, the author appears to be linked with various special interests opposed to this kind of support for renewable energy (Johnson, 2009). The facts on Spain clearly show a different picture of development.

This again underlines the point that vested interests will continually seek to undermine confidence in approaches they are threatened by. They use the well-proven tactic of FUD (fear, uncertainty and doubt) that was so successfully used by industry to oppose bans on smoking and limitations on global warming gases. This tactic is unlikely to be given up, but it does allow the exercise of research and communications by those in favour of renewable energy and/or FITs. It can also serve to publicly highlight the tactics of opponents.

10.2.5 Find a political sponsor and engage the political process

When a campaign for FITs reaches a certain point, it will become time to draft legislation and participate in the political process. This component is perhaps the hardest part, for it requires patience, tolerance, diplomacy, tact and often directness, strength and the single-mindedness of a bull elk in the rut. Depending on the nation one hails from, there may be a decision to be made on the scale best targeted for legislation – national, state, provincial or municipal. North America, India and Australia could have any of the above, for example.

Finding a political sponsor is one of the most important parts of the process. If in government, lawmakers can push their own party to adopt the legislation, or if in opposition, a more general call can be made for support for a bill or amendment. Nations have differing parliamentary systems and methods for introducing legislation or just raising the subject itself. For example, the UK has the Private Members Bill, the Ten-Minute Rule, the Early Day Motion and amendments to existing bills.

Identifying and working with a top parliamentary lobbyist can make things much easier and improve chances of success. This person should ideally have a good track record, be known and trusted, and know the political system inside out. They will be able to guide the coalition and advise them on political strategy within and around parliaments and legislative bodies. The main focus should be the building of support for the FIT, ideally across all parties. Parliamentary events aimed at open discussion can help achieve some of this, but coalitions will need to work on various fronts. Letter writing to the relevant minister or secretary of state can help describe the intentions and arguments of the coalition. Open letters of this kind can then be made public to increase awareness of the coalition, its focus and its arguments. Having a few informed journalists involved, who are willing and able to keep the government and other interests honest in the national press, will be especially helpful.

It is worth noting that non-profit organizations may not be able to do direct lobbying. The laws governing their activity differ, but often exclude 'lobbying', and this may need to be carried out on by other members of the coalition. They can do 'educational' work, however, which is general awareness-raising, policy research, and community outreach. They can also facilitate contact between various agents, including industry, policy experts, manufacturers and so on.

Coalitions working very closely with the government may find themselves compromised and less able to publicly criticize and pressure the government. However, NGOs within the coalition may be only too happy to publicly apply such pressure – as is often their mandate. Those familiar with lobbying and the political process will need no further information here, and circumstances will differ significantly based on each country. As a general rule, the wider and deeper the coalition, the more refined and accommodating of varied concerns it, and the ensuing legislation, will be.

As for the public, energy is so fundamental to all our lives, yet so remote from the experience of most citizens, that it is a difficult thing to communicate. The arguments that tend to work are those which offer people something tangible, such as jobs or a guaranteed profit from investment. These are the same things that the government will want to know about. They will also want to know several key things: how much will it cost? Who pays? Who supports this? These questions will generally follow a discussion of the benefits of FITs, and precede a discussion of why one is not already in place, and who opposes it.

With a determined group of cross-party policy makers and a few informed journalists on board, a proactive and decently funded coalition can achieve a great deal when they follow a strategy, carefully research and argue, expose self-interested opposition, and stay the course.

The arguments in favour of renewable energy means the technologies practically sell themselves. What remains is for enthusiastic, professional and organized coalitions and independent voices to create momentum towards the goal of increasingly rapid, democratized deployment of renewable energy.

10.2.6 Design and implementation

But if the government agrees, what then? Once a government commits to introduction of legislation, the job is still not done. Coalition members should present clear proposals for all areas of the legislation, helping to determine eligible technologies, sizes, appropriate tariff levels and so on. From observation of activity in various countries, tariff levels are often set quite well, but one may be inclined to push the government to begin towards the higher end of the proposals. If the tariff is set too low and creates no deployment, nothing is learnt in the process, but if they are set higher and take-up is swift, there is more to be learnt on appropriate level of tariff reduction. This does not suggest setting the highest possible tariff, but a level which experts broadly agree will be sufficient to stimulate deployment. This really comes back to the central issue around the tariffs, which is that they must strike a balance between investment security and end-user costs. Investors will always want high returns, but the public (and the politicians that represent them) want lower and stable electricity prices.

To policy makers in the current economic climate, the policy will be sold mainly on the basis of job creation and economic stimulation. Secondary benefits will be energy security and emissions reductions. Tertiary benefits will be the social and environmental consequences. This should not leave the green-hearted downhearted. The call for a fundamental shift in values and ethics is not pointless, but we must be realistic in our assessment of what drives policy makers and public support. A variety of arguments is useful, and demonstrates just how transformative this policy can be, if designed and implemented well, according to local conditions.

10.3 Final Thoughts

Otto Von Bismarck once said that the making of laws was like the making of sausages: it's often better not to know how they are produced. His comment underscores that the process of making legislation of any type is often conflicted and chaotic. FITs are no different, as it can be a long, hard road to get a decent FIT implemented, especially when you push for a law and not just a ministerial order (see Section 4.9). As a highly experienced and successful FIT campaigner reminded us, it is worth underlining the need for patience. He described the process of achieving major policy change as 'messy and frustrating'. The implementation and enactment of FITs does not happen quickly, and sometimes making compromises is an uncomfortable but inevitable ingredient of effective decision making among disparate interests.

However, readers must not forget that few policies and laws bring the types of benefits that FITs do. They create a democratization of energy production, making everyone a potential energy producer or investor. They empower citizens and communities in a new way, and pave the way for improved green legislation. This is radical, but is arguably part of the most important and unavoidable transition in human history – from energy sources which are driving a mass extinction unprecedented in the Earth's history, to an energy system based on benign renewable cycles, the engagement of ordinary people and a lucrative green economy.

REFERENCES

Johnson, K. (2009) 'Green jobs, ole: Is the Spanish clean-energy push a cautionary tale?', http://blogs.wsj.com/environmentalcapital/2009/03/30/green-jobs-ole-is-the-spanish-clean-energy-push-a-cautionary-tale/

Lacey, S. (2008) 'US state solar debate: Will SRECs create unhealthy market concentration?', Renewable Energy World.com, www.renewableenergyworld.com/rea/news/article/2008/05/u-s-state-solar-debate-will-srecs-create-unhealthy-market-concentration-52339

Lappé, F. M. (2007) *Getting a Grip: Clarity, Creativity and Courage in a World Gone Mad*, Small Planet Media, Cambridge, MA

Mendonça, M. (2007) *FITs: Accelerating the Deployment of Renewable Energy*, Earthscan, London

Mendonça, M., Lacey, S. and Hvelplund, F. (2009) 'Stability, participation and transparency in renewable energy policy: Lessons from Denmark and the United States', *Policy and Society*, vol 27, pp379–398

Sovacool, B. K. (2008) '*The Dirty Energy Dilemma: What's Blocking Clean Power in the United States*', Praeger, Westport, CN

Toke, D. (2005) *Wind Power Outcomes: Myths and Reality*, paper for the 11th Annual International Sustainable Energy Research Conference in Helsinki, 6–8 June

Notes

Chapter 3

1 This section is largely based on Jacobs and Pfeiffer, 2009.

Chapter 4

1 The annual average insolation (kWh/m^2) is 1825 in Spain and 1095 in Germany.

Chapter 6

1 The number of countries in Table 6.1 differs from the other current reports as we are only showing countries which are currently supporting renewable energies under this support mechanism.
2 In addition, there were minor revisions in 2003 and 2006.
3 *Renewable Electricity Tariffs ('Feed-in Tariffs for Small Scale Generation of Electricity') – Renewable Heat Tariffs ('Renewable Heat Incentive'): Preliminary Recommendations on Their Implementation from the Renewable Energy – Output from working groups and industry input co-ordinated by the Renewable Energy Association (REA)*, 26 March 2009, www.r-e-a.net/document-library/policy/policy-briefings/RET_Report1-1.pdf.
4 For excellent summaries of PURPA, see Paul J. Joskow (1979) 'Public Utilities Regulatory Policy Act of 1978: Electric utility rate reform', *Natural Resources Journal*, vol 19 (1979), pp787–810; Richard F. Hirsh (1989) *Technology and Transformation in the American Electric Utility Industry*, Cambridge University Press, Cambridge; Richard F. Hirsh (1999) *Power Loss: The Origins of Deregulation and Restructuring in the American Electric Utility System*, MIT Press, Cambridge, MA.
5 The different price for electricity generation from diesel generators in Nairobi and Mombasa is related to the additional transport costs (Mombasa is located on the coast).
6 See www.mnre.gov.in.

Chapter 7

1. Interview with Paul Gilman, Oak Ridge Center for Advanced Studies, Oak Ridge, TN, 3 August 2005.
2. Interview with Ralph Loomis, Exelon Corporation, Chicago, Illinois, 6 October 2005.
3. Interview with Thomas Chrometzka, German Solar Industry Association (BSW-Solar), Berlin, Germany, 28 July 2008.

Chapter 8

1. This categorization of barriers is based on a framework developed in Sovacool, B. K. (2008) *The Dirty Energy Dilemma: What's Blocking Clean Power in the United States?*, Praeger, Westport, CT, pp123–200.
2. Research interview with Wilson Prichett, 16 February 2006, Blacksburg, Virginia.

Chapter 9

1. This story is fondly taken from Benjamin K. Sovacool, (2008a) 'Matter of stability and equity: The case for federal action on renewable portfolio standards in the US', *Energy and Environment*, vol 19, no 2, pp241–261.
2. Those wishing to read more about the genesis of RPS as a policy tool should explore American Wind Energy Association (1997) *The Renewables Portfolio Standard: How it Works and Why it's Needed*, American Wind Energy Association, Washington, DC; Cory, K. S. and Swezey, B. G. (2007) *Renewable Portfolio Standards in the States: Balancing Goals and Implementation Strategies*, National Renewable Energy Laboratory Technical Report NREL/TP-670-41409, Golden, CO; Rader, N. (1997) *The Mechanics of a Renewables Portfolio Standard Applied at the Federal Level*, American Wind Energy Association, Washington, DC; Rader, N. (2003) 'The hazards of implementing renewables portfolio standards', *Energy and Environment*, vol 11, no 4, pp391–405; Rader, N. and Hempling, S. (2001) *The Renewables Portfolio Standard: A Practical Guide*, National Association of Regulatory Utility Commissioners, Washington, DC; Rader, N. and Norgaard, R. (1996) 'Efficiency and sustainability in restructured electricity markets: The renewables portfolio standard', *Electricity Journal*, pp37–49.
3. The Renewables Obligation in the UK is trying to tackle this problem through the implementation of 'banding', i.e. not all renewable electricity producers receive the same amount of certificate per unit of electricity. More costly technologies will receive more certificates. This way, the Renewables Obligation will become a technology-specific support instrument. However, the complexity of the support mechanism and the administrative costs will increase.
4. David Elliott, Interview with author, London, United Kingdom, 9 November 2007.
5. Till Stenzil, interview with author, London, United Kingdom, 16 November 2007.
6. Senior government official in the United States, interview with author, 2 August 2008.

7 California Bill, AB 1890, Article 7, Research, Environmental, and Low-Income Funds, Section 381(c)(1) to (c)(3). The law also mandated utility funding of low-income programmes at 1996 levels or higher. See also Wiser, R., Pickle, S. and Goldman, C. (1996) *California Renewable Energy Policy and Implementation Issues: An Overview of Recent Regulatory and Legislative Action*, Report LBNL-39247, UC-1321. Recommendations on how to allocate funds for renewable technologies are included in California Energy Commission, Renewables Program Committee (undated) *Policy Report on AB 1890, Renewables Funding*. This bears no date, but an accompanying letter was dated 7 March 1997.

Index

additionality 72, 160
administrative barriers 80, 85, 136
administrative procedures 34–35
  costs of 69, 71
advanced FIT design 39–55
  and demand *see* demand-oriented tariff differentiation
  forecast obligation in 39, 41, 44–46
  inflation-indexed tariffs in 39, 52–53
  innovative features payment 39, 53–55
  and limited/total generation 39, 48
  location-specific design 39, 47–48
  and market integration 39
  and national conditions 40
  and premium FITs *see* premium FITs
  special tariff payments in 39
  spot market participation 39, 42–43, 44
  tariff degression in 39, 49–51, 64
AEE (Spanish wind energy association) 87
aesthetics 130, 132, 140, 145
Africa 78, 79, 102
  see also specific countries
Age of Stupid (documentary) 11
agricultural waste 86, 102, 118
air quality/pollution 62, 67, 141
Algeria 78
Allen, Anthony 134
alternative fuels 3, 7, 8, 119
American Solar Energy Society 7–8
animal waste 17, 26
annuity method 21–22
Apollo Alliance 2, 6
application process 70
Argentina 40, 57, 68, 78, 102
Arizona (US) 123, 151, 155
Asia 4, 78, 79, 106
  see also specific countries

Australia 4, 78, 79, 150, 151, 162, 192
  national FIT scheme in 96, 97
  state FIT schemes in 96–102
  targets in 96
Australian Capital Territory 97–98
Austria 17, 34, 55, 78, 136, 158
avian mortality 140–141, 142, 143
avoided costs 19, 62, 80, 93

bad FIT design 34, 57–64, 77
  capacity caps 59, 63–64, 68
  financing mechanisms 61
  flat-rate tariff 59–60
  legal/legislative issues 64
  maximum/minimum tariffs 60, 104
  purchase obligation exemptions 60–61
  tariff calculation methodologies 61–63, 193
  tariffs too high 58–59, 62
  tariffs too low 57–58, 62
bagasse 102, 105
BANANA (build absolutely nothing anywhere near anything) 145
barriers to renewable energy 129–146
  aesthetic/environmental 130, 140–145
  cultural/behavioural 130, 138–140
  financial/market 130–134
  FITs and 131, 132
  information failure 130–131
  political/regulatory 130, 134–138
  vested interests 129, 133–134, 185–186, 191
base-load power 111, 115, 118, 122
batteries 3, 48, 121, 126
BEE (Renewable Energy Federation, Germany) 44
Belgium 151, 156, 164, 177
Berendt, Christopher 159
best practice 6, 19, 57, 77, 79

bidding system *see* tendering
biofuels 3, 7, 8, 119
biogas 22, 26, 43, 86, 92, 140
  methanization of 53
biomass 9, 16, 43, 44, 49, 82
  costs 26, 140, 144
  crops 17, 26, 118, 141
  eligibility for FIT 17
  FIT calculation for 21, 22, 63
  and mini-grids 73
  negative impact of 67
  potential of 5
  pulp/paper 105
  reliability of 111–112, 115
  tariff payment duration for 86–87
  thermochemical gasification of 53
BIPV (building-integrated PV) 8, 17, 27, 42, 54–55, 82, 86, 92
birds *see* avian mortality
Blue Green Alliance 6
bonus-malus system 46
Brazil 9, 68, 78, 102, 176
Britain 4, 78–79, 87–90, 169, 192
  energy oligopoly in 88
  energy prices in 88
  FIT tariff calculation in 89
  fuel poverty in 87–88, 89
  heat tariff in 87, 89
  legislation in 87, 88–89
  nuclear power in 88, 89
  REA working groups in 88–89, 90
  RO (Renewables Obligation) in 88, 177, 178
  RPS in 150, 151, 152–153, 154
  small-scale schemes in 87, 88
  targets in 89
  wind power in 116, 125
budgets, national 61, 70
building-integrated PV *see* BIPV
building renovation 3
building standards/codes 8, 35
burden-sharing 28–29, 31–34, 68–69
bureaucracy 136
Bush, George H.W. 135
Bush, George W. 2

CAES (compressed air energy storage) 122–123
California (US) 93, 94, 115, 116–117, 120–121, 122
  non-FIT schemes in 150, 151, 159, 162, 165
Canada 4, 9, 78, 90–92, 149, 151
  grid bottleneck in 31
  nuclear power in 113

provincial energy policy in 90
  see also Ontario
capacity-building 2–4
capacity caps 59, 63–64, 80, 87, 155
  in developing countries 68–70, 76
  in US 93–94
Cape Wind project (Massachusetts, US) 145, 155
capital investment 62, 131–132
carbon market 71–72
carbon sequestration 3
CCS (carbon capture and storage) 89
CDM (Clean Development Mechanism) 68, 71–72, 76
  and additionality 72
centralization of power 114
CER (Certified Emission Reduction) units 71
certificates *see* tradable certificate schemes
Chabot, Bernard 47
China 4, 9, 68, 102, 106, 151, 176
Clean Development Mechanism *see* CDM
climate change 3, 9, 11, 62, 64, 68, 108
Clinton, Bill 6
COAG (Council of Australian Governments) 97, 98
coal/coal plants 84, 89, 91, 104
  'clean' 153, 185
  external costs of 141, 142, 144
  subsidies for 137
  unreliability of 112
Colorado (US) 7, 117, 120, 151, 164
combined power plants 44, 118–120
comfort factor 139–140
communities
  green technologies in 3, 10–12, 42
  rural 72–76, 96, 145
competitiveness 124–125, 175–176, 178
compressed air energy storage (CAES) 122–123
consumption patterns 138–140
conventional energy industry 10, 28, 30, 39
  and avian mortality 141, 142, 143
  and avoided external costs 62
  as barrier to renewables 129, 133–134, 185–186, 191
  and capacity caps 64
  cost overruns in 113, 114
  and forecasts 44, 113–114
  'lumpy' 113
  misinformation by 185–186, 191
  opposition to FITs 185–186
  power outages in 112–113
  reserve capacity in 113
  and subsidies 136–137
  unreliability of 112–115, 126

corruption 48, 70
cost–benefit analysis 81–85
cost-covering remuneration approach 19, 20–22
cost recovery 52
cost-sharing 28–29, 31–34
Council of Australian Governments *see* COAG
Croatia 78
CSP (concentrated solar power) 5, 105
  and FIT calculations 24, 25
Cuba 119–120
Cyprus 78
Czech Republic 40, 78, 164

dams 140
DBU (Federal Foundation for the Environment, Germany) 15, 20
debt-equity ratio 24
decentralization 18, 136
decommissioning costs 20
deep connection charging 32–33
Delaware (US) 151, 158
demand and merit-order effect 83–84
demand-oriented tariff differentiation 30, 39, 43
democracy 12, 19, 178, 186–188
  living/innovative 186–187
Denholm, Paul 122
Denmark 10–11, 33, 53, 55, 77, 78, 91, 176
  barriers to renewables in 135
  non-FIT schemes in 153, 158, 164
  premium FITs in 40
  wind power in 116, 125, 135
dependency creep 11
developing countries *see* emerging economies
de Vos, Rolf 157
Dinica, Valentina 153
discount rate 131
diversity principle 186
double metering 164
DSO (distribution system operator) 29, 74, 75

eco-labelling 8
economic stimulus packages 4
economies of scale 49, 50, 63, 68
Ecuador 78
education, energy 12, 139, 188–189
EEG (Renewable Energy Sources Act 2000, Germany) 30, 33, 80–81
  amendments to (2004, 2009) 80
  progress report 82
electricity exports/imports 18
electricity generation costs 21–22
electricity grids 3, 16, 20, 30–34
  bottleneck in 31
  connection to *see* grid connection
  in developing countries 73
  and forecast obligations 45–46
  and special tariff payments 39, 46
  see also transmission lines
electricity price 12, 19, 28, 41, 77
  in emerging economies 68
  and merit-order effect 83–84
  and premium FITs 41–42
  and tariff levels 63, 80
  uncertainty of 152, 154
Elliott, David 152–153
emerging economies, FIT in 20, 24–25, 67–76
  application process 70
  capacity caps 68–69, 76
  and CDM 68, 71–72, 76
  cost distribution 71
  financing mechanisms for 61, 64, 74–75
  FIT fund 68, 70–71, 76
  inflation in 68, 105
  international donors and 70, 74
  for mini-grids *see* mini-grids
  and national planning 69
  potential for 67–68
  role of regulator 68, 70, 74, 75
employment 2, 84–85
  see also green jobs
energy audits 91
energy crops 17, 26, 118, 141
energy development agencies 75
energy efficiency 7, 8, 71, 105, 120
energy independence 3, 62, 84, 119
energy-intensive industries 39, 55
energy markets 9, 10, 52, 129
  European 63
  growth in 50–51, 63–64, 175
  integration of 15, 39
  intra-day 41, 45, 67
  and monopolies/oligopolies 67–68, 133–134
  see also spot market
energy policies, and monopolies/oligopolies 10, 133–134
Energy Savings Trust (UK) 169
energy security 62, 114
entrepreneurs 135–136
equal opportunities 7
Estonia 40, 59, 60, 78, 158
EU ETS (EU Emissions Trading Scheme) 89
Europe 77, 78, 91, 163
  see also specific countries
European Emissions Trading Scheme 83
European Photovoltaic Policy Platform 26
European Union (EU) 4, 79, 149, 176

administrative recommendations of 34
barriers to renewables in 136
biomass defined by 17
Directive for Renewable Energies (EC) 35
Guarantees of Origin in 156
and mini-grids 73
non-FIT support schemes in 150–151, 156, 176, 177
targets 89
wind farm reliability in 117–118
externalities 62, 140–145

fact-finding trips 190
families 7, 138–139
Fell, Hans-Josef 18
financing mechanisms for FITs 29–30
Finland 40, 79, 116, 125
firm/non-firm power generation 103
first-come, first-served system 69
FIT fund 68, 70–71, 76
FITs (feed-in tariffs) 8
  benefits of 12
  fixing level of 57–60
  flexibility of 44
  history/development of 77–78
  opposition to 9–10, 78, 185–186, 191
  and other support schemes 150, 176–179
FITs campaigning 185–194
  allies in 189–190
  and communications 191
  and democracy/participation 186–188
  and education/information 188–189
  and energy industry opposition 185–186, 191
  fund-raising for 190–191
  and policy analysis 186–187
  and policy design/implementation 193
  sponsors/coalitions for 191–193
  and workshops/fact-finding 190
FIT designs 15–64
  administrative procedures 34–35
  advanced *see* advanced FIT design
  bad *see* bad FIT design
  basic 15–37
  basic, checklist for 36–37
  complexity/simplicity in 15, 34–35
  duration of payment 27–28, 81, 86–87, 92, 95, 97–101, 103, 105, 106
  eligible plants for 17–19
  for emerging economies *see* emerging economies
  financing mechanisms for 16, 29–30
  progress reports in 36

purchase obligations 16, 29–30
size-specific tarrifs 18, 26–27
and spatial planning 35
tariff calculation for *see* tariff calculation
technology-specific 16–17, 26
flat-rate tariff 59–60
flexible tariff degression 39, 64
Florida (US) 90, 93–94
forecast obligation 39, 41, 44–46
forestry products 17, 26, 86
fossil-based power 28
  unreliability of 112
France 4, 9, 47, 48, 53, 64, 78, 113, 158, 175, 177
  ADEME 23
  avoided costs in 62
  bureaucracy in 136
  FIT tariff calculation in 23–24
  location-specific tariffs in 47–48
fraud/corruption 48, 70
free riders 72
fuel cells 7, 8, 89, 153
fuel costs 20, 22
fuel poverty 87–88, 89
full load hours 44, 52
fund-raising 190–191

G-20 countries 1–2
Gainesville (Florida, US) 90, 93–94
gasification 53, 89
GEA (Green Energy Act, Ontario) 91
geothermal power 9, 16, 43, 49, 82, 102–103
  and air pollution 141
  external costs of 141, 144
  FIT calculation for 21
  in hybrid systems 118
  potential of 4–5
  reliability of 111–112, 115
  tariff payment duration for 86
Germany 4, 5, 18, 64, 77, 78, 79, 80–85, 91, 113, 190
  barriers to renewables in 130–131, 186
  BIPV in 55
  capacity cap in 80
  combined technologies scheme in 44
  cost–benefit analysis of FIT in 81–85
  DBU 15, 20
  dispersed PV units in 117
  electricity price in 63, 83–84
  energy-intensive industries in 55
  FIT complexity in 15
  FIT reports in 36
  FIT target in 35, 81

FIT tariff calculation in 19, 20–22, 57–58, 63, 80
forecast obligation in 45–46
green jobs in 7, 9
grid connection in 33
hybrid systems in 118–119
hydropower in 80
innovative features payment in 53, 54, 55
non-FIT schemes in 153, 167, 176, 177, 178
premium FITs in 40, 42
priority grid access in 30, 31
Renewable Energy Federation (BEE) 44
renewable energy growth in 80–81, 82
Renewable Energy Sources Act (2000) *see* EEG
tariff degression in 49, 50–51, 81
tariff increases in 52, 82–83
voltage dips in 46
wind power in 53, 54, 80, 116, 125, 131, 187
Ghana 68, 102
globalization 58
GOs (Guarantees of Origin) 155, 156
grants 3
Greece 78
green economy 1–12
  capacity-building for 2–4
  investment potential of 4–5, 7
  opposition to 9–10, 11, 78, 88, 145, 185–186, 191
  public support for 10–12
  three principles of 7
  use of term 1
Green For All 2, 6
greenhouse gases 62, 71, 72, 84, 140
green jobs 3, 5–9, 12, 84–85, 191
  common features of 7
  defined 6–7
  policy drivers for 7–9
  'shades of' 6
Green Party (Germany) 51
green power programmes 150, 157, 161–164
  participants in 162, 163
  strengths of 162
  weaknesses of 162–164
green procurement 8
green tags *see* RECs
green technologies
  combining 44
  community involvement in 3, 10–12
  definitions of 17
  and FITs 16–17, 26
'grey' electricity *see* conventional electricity industry
grid connection 30–34, 80

access issues 30–31, 68, 85
cost-sharing 29, 31–34
deep/shallow/super-shallow 32
and dominant players 68, 133–134
DSO/TSO 29
grid integration 111, 112, 116–118, 124–125
Guarantees of Origin (GOs) 155, 156

Harvard Business School 131, 132
heat tariff 87, 89
Heffernan, Patrick 6
Holst, Henning 187
HOME Investment Partnerships (US) 3
homeowners 131–133
  see also families
hot dry rock technology 53, 86
house-building/renovating 3, 131–132
households 7, 138–139
Hungary 43, 59, 78, 136
Hvelplund, Frede 187
hybrid systems 118–121
hybrid transport 8
hydropower 9, 16, 43, 44, 49, 82, 92, 102–103, 105
  external costs of 140, 144
  and FIT calculations 21, 25, 63
  FIT size limit for 18
  lead times for 34
  and mini-grids 73
  pumped hydro systems 121–122
  reliability of 111–112, 115
  tariff payment duration for 86

independent power producer (IPP) 74
India 9, 78, 106–108, 151, 176, 192
  hydropower in 106
  legislation in 106, 107, 108
  renewable energy target in 107
  solar power in 107–108
  state schemes in 107–108
  wind energy in 106, 107
Indonesia 78
industrialized countries 70, 74, 137
  burden-sharing by 68–69, 76
  carbon trading by 71
industry, energy-intensive 39, 55, 133
inflation 21, 22, 23, 57
inflation-indexed tariffs 39, 52–53, 68
information/misinformation 112, 130–131, 139, 146, 185, 188–189, 191
infrastructure investment 3, 149
innovation 39, 53–55, 129
innovative democracy 187

Inslee, Jay 95–96
interconnection *see* grid integration
intermittency *see* reliability
International Emissions Trading Scheme 71
International Energy Agency (IEA) 115, 126, 167, 176
international policy diffusion 78–79, 91
intra-day markets 41, 45, 67
inverter technology 114
investment 20, 153
  capital 62, 131–132
  costs 22
  grants 79
  potential 4–5, 7–8, 10–11
  private-sector 173
  rate of return on *see* rate of return
  security 15, 29–30, 61, 85, 176
IPP (independent power producer) 74
Ireland 34, 53, 78, 125
IRR (internal rate of return) 23, 25, 59
Israel 68, 78, 102
Italy 4, 40, 78, 151, 177
  BIPV in 54–55
  TGC in 79
ITCs (investment tax credits) 170–172

Jacobs, David 15
Japan 4, 9, 106, 150, 151, 164, 167
JI (Joint Implementation) 71
Jones, Van 2, 6, 7

Kenya 60–61, 68, 69, 71, 78, 102–104
  hydropower/geothermal in 102, 103
Kingston (Ontario, Canada) 91
Knight, R.C. 32
knowledge transfer 112
Kombikraftwerk 118–119
Koomey, J. G. 131
Kyoto Protocol 2, 71, 72

landfill gas 8, 17, 49, 82, 86, 105, 118
  and FIT calculations 21, 24, 25, 28, 92
landlords/tenants 132–133
landscapes 11, 36, 145
  see also visual impact issues
land use 118, 122, 125, 136, 140, 141
Langeraar, J.-W. 157
Lappé, Frances Moore 186–187
Latvia 78
Lawrence Berkeley National Laboratory 160, 172, 174
lead times 25, 34, 113, 114, 126, 136, 175
levelized electricity cost approach 24–25

licensing process 20, 34, 60, 136, 138
liquidity 41, 67
Lithuania 78
living democracy 186–187
lobbying 191–193
local government 3
location-specific design 39, 47–48
Lovins, Amory 185
low-income housing 3
Lund, Peter 129
Luxembourg 78

McCain, John 6
Macedonia 78
Malta 78
maps, resource 16
market-based mechanisms 9, 10
market conditions 52
market growth 50–51, 63–64, 175
market integration 15, 39
Martin, T. 89
Maryland (US) 151, 158, 162
Massachusetts (US) 145, 155, 158, 167–168
materialism 1
Mauritius 78, 102
maximum/minimum tariffs 60, 104
merit-order effect 83–84
methanization 53
Mexico 120, 164
micro-CHP 89
Miliband, Ed 11, 88
military and green economy 3, 189–190
mine gas 21, 30, 33, 45, 49, 82
mini-grids 72–76
  backup power for 73
  described 72–73
  income sources 74, 75, 76
  liberalized approach to 74
  monopolist approach to 74–75
  stakeholders in 73–74
Minnesota (US) 93–94, 116–117
misinformation *see* information
mobile phones 11
molten salt plants 115, 123–124
monopolies/oligopolies 10, 67–68, 106, 133–134, 154–155
  RESCOs 74–75

Nakarado, G. L. 131
NASA 3, 189
natural gas 14, 84, 119, 142, 153
NERSA (National Energy Regulator of South Africa) 104, 105

net/gross FIT schemes 96–97
Netherlands 40, 55, 78
net metering 120, 150, 164–166
　benefits of 164–165
　drawbacks of 165–166
net present value method 22
Nevada (US) 151, 155
New England (US) 131, 153, 158
New Jersey (US) 94, 151, 165
New South Wales (Australia) 98
New York (US) 116–117, 151, 153, 158, 162
New Zealand 96, 170
NFFO (Non-Fossil Fuel Obligation) 174–175
NGOs (non-governmental organizations) 4, 55, 88, 108, 191
Nicaragua 78
Nigeria 68, 102
NIMBY (not in my backyard) 145
North America 78, 79, 90, 91, 164, 191, 192
　see also Canada; United States
Northern Territory (Australia) 98–99
Norway 71, 115, 116, 125, 156, 169–170
nuclear power 11, 28, 30, 88, 89, 104
　external costs of 141, 142, 143, 144
　phasing out 5
　subsidies for 137–138
　unreliability of 112, 113–114

O&M (operation and maintenance) 20, 22, 23, 25, 75
Obama, Barack 2, 6
ocean current energy 86
OECD countries 137
offshore wind turbines 17–18
oil crisis 77
oil price 103, 104
Ontario (Canada) 31, 35, 90–92, 113, 151
　grid access in 90
　legislation in 91
　Power Authority (OPA) 90, 91, 92
　RESOP in 90
　wind power in 117
operation and maintenance *see* O&M
outages 104, 112–113, 114, 117
ownership issues 40–41

Pacific Gas and Electric Company 115
Pakistan 68, 78, 102
Pasqualetti, M.J. 145
pass-through costs 71
patents 129
peak/off-peak demand 43, 112
permits *see* licensing

personal freedom 139–140
personnel costs 21–22
Philippines 78
photovoltaic systems *see* PV
PI (profitability index) 24–25
Pitt, Damian 136
planning process 55, 136
plug-in hybrid vehicles 121
Poland 151, 158
policy diffusion, international 78–79
political issues 134–138, 146
　electricity price 68
　financing mechanisms 61
　FIT campaigning 185–194
　government funding 70–71
　legal status of FITs 64
　tariff levels 57, 58, 75
population 11, 47
Portugal 9, 34, 48, 62, 77, 78, 125, 176
poverty reduction 67
PPAs (power purchase agreements) 74, 105, 158
pre-capitalization 89
premium FITs 10, 39, 40–42, 86–87
　benefits/drawbacks of 42
　cap/floor in 40
　and fixed FITs 42
　prerequisites for 40–41, 67
price fixing 9–10
Prichett, Wilson 134
principal-agent problem 132–133
procurement, green 8
producer responsibility 8
production capacity 48
profitability 21, 59, 74, 75, 135–136
profitability index (PI) 23–24, 47
PTCs (production tax credits) 172–174, 175
public housing 3, 133
public support 10–12, 17, 71, 97, 111
public utilities 19, 42, 74, 133–134, 153
pumped hydro systems 121–122
purchase obligations 16, 29–30
PURPA (Public Utility Regulatory Bodies Act, US, 1978) 59, 77, 93
'Put People First' coalition 2
PV (photovoltaics) 9, 17, 43, 59, 82, 92
　in combined/hybrid systems 116–121
　costs 26
　FIT calculation for 21, 63, 94–95
　location of 47
　market growth in 64
　and mini-grids 73
　potential of 5, 7
　reliability of 114, 115

rooftop installations *see* BIPV
tariff degression for 49, 87
tariff payment duration for 86–87
and TGC 79
variability of 30, 112
see also solar panels
pyrolysis 89

Queensland (Australia) 99
quotas *see* RPS

R&D (research and development) 3, 5, 63, 149, 150, 166–167, 172, 178
eight areas of 166
global expenditure on 168
railways 3
Rankine cycles 53
rapid transit systems 8
rate of return 19, 24–25, 41–42
see also IRR
raw materials 52
Reagan, Ronald 134
REA (Renewable Energy Association, UK) 88–89, 90
RECs (renewable energy credits) 155–160, 178
bundled/unbundled 156
complexity of 157–158
drawbacks of 157–160
features of 156–157
and FITs 176–177
GOs 155, 156
price fluctuations of 158–160
see also TGCs
recycling 7, 8
registry 70
REGOs (Renewable Energy Guarantee of Origin Certificates) 156–157, 158, 161
regulator 68, 70, 74, 75
reliability 111–123
backup technology 121–123
of conventional power plants 112–115
and energy use changes 112
of hybrid systems 118–121
and market 41
of non-intermittent renewables 115
and small-scale systems 114–115
of wind/solar systems 116–118
renewable energy
100 per cent 63
barriers to *see* barriers to renewable energy
external costs of 140–141, 143, 144
factors in success of 9
fluctuations in *see* reliability

potential of 4–5, 7, 8, 121
see also specific forms of renewable energy
Renewable Energy Association *see* REA
renewable energy credits *see* RECs
rented property 132–133
RESCOs (rural energy service companies) 74–75
research and development *see* R&D
RESOP (Renewable Energy Standard Offer Program, Ontario) 90
retroactive tariffs 89
Rhode Island (US) 94, 151, 158
Ringo, Jerome 2
risk 17, 28, 41–42, 58, 62, 70, 132, 135–136, 176
Rocky Mountain Institute 120
ROCs (Renewable Obligation Certificates) 152, 155, 177–178
RPOs (Renewable Purchase Obligations) 106–107
RPS (renewable portfolio standards) 8, 26, 61, 150–155, 157, 173
benefits of 151–152
described 150
drawbacks of 152–153
and FITs 176
FITs superior to 176
growth of 150–151
rural communities 72–76, 96, 145

salt storage 115, 123–124
sanitation 8
SBCs (system benefits charges) 150, 167–170, 173
benefits/drawbacks of 170
development of 169
scale/size issues 34, 42, 69, 70, 71–72, 79, 87, 96, 153, 157–158, 174
and reliability 114–115, 116
see also capacity caps; mini-grids
schools, modernizing 3
Schwäbisch Hall power plant 118–119
security of supply 62, 114
sewage gas 17, 26, 49, 82, 89
shallow connection charging 33
size-specific tariffs 18, 26–27
Slovakia 60, 78
Slovenia 40, 78
Smitherman, George 90–91
Smith, J. C. 125
socio-economic development 67, 145
solar maps 16
solar panels 8, 120, 144
hazardous substances in 140

payback from 132
reliability of 114
thin film modules 55, 91
solar power
potential of 4–5
scale/size issues with 69
and spatial planning 35
Solar Renewable Energy Certificates (SRECs) 10
solar thermal technology 16, 89, 144
reliability of 111–112, 115
solar tracking devices 48, 120
South Africa 68, 71, 78, 102, 104–105, 176
FIT tariff calculation in 24–25, 27, 36, 72
South Australia 99–100
South Korea 4, 61, 78, 106
Spain 9, 34, 64, 77, 78, 79, 85–87, 91, 123, 176, 190, 191
capacity cap in 87
electricity budget deficit in 61
electricity price in 83
FIT legislation in 64, 85, 87
FIT reports in 36
FIT tariff calculation in 19, 20, 58–59, 63, 85
forecast obligation in 44–45
increasing tariffs in 52
inflation indexation in 53
investment security in 85
premium FIT in 40, 41, 42, 86–87
PV in 51, 58–59, 61, 64, 69, 86
renewable energy growth/targets in 87
size-specific FIT in 18
small-scale schemes in 87
spot market in 42–43
targets in 87
tariff degression in 50, 51, 87
tariff payment duration in 86–87
voltage dips in 46
wind power in 52, 53, 54, 86, 87, 125
spatial planning 34, 35
SPEED (Sustainably Priced Energy Enterprise Development) Program (US) 95
split incentives 132–133
spot market 28, 41, 42–43, 44, 67, 83
SRECs (Solar Renewable Energy Certificates) 10
Sri Lanka 78
standard offer contracts 93
Stavros, Richard 134
Stern Review (2006) 9, 178
'stop and go' cycles 64, 69
storage technology 121–123
see also batteries
stranded costs 134
subsidies 136–138

super-shallow connection charging 33
supply/demand 43, 112
Sweden 113, 116, 125, 151
Switzerland 78, 113, 156

Taiwan 106, 151
targets 8, 9, 35–36, 81, 149, 150, 155
tariff calculation 19–26, 57–63, 193
bad methodologies for 61–63, 193
cost-covering remuneration approach 19, 20–22
online tool for 26
profitability index method 24–25
reasonable rate of return approach 19, 20
tariff degression 39, 49–51, 64, 81
Tasmania 100
taxation 61
tax credits 8, 135, 149, 150, 170–174, 178
investment (ITCs) 170–172
production (PTCs) 172–174, 175
tax incentives 79
technology development 68–69, 80
technology differentiation 17, 26, 59–60, 61, 80
telecommunications 11
tendering 150, 174–176, 178
problems with 175–176
Tennessee Valley Authority 93
Texas (US) 151, 153
TGCs (Tradable Green Certificates) 10, 26, 67, 71–72, 79, 153, 155, 178
see also RECs
Thailand 78, 164
thermochemical gasification 53
Toke, David 174
tradable certificate schemes *see* TGC
trade unions/associations 4, 189, 190
traffic congestion 3, 140
transaction costs 129
transmission costs 125
transmission lines 18, 29, 32, 33, 112, 115, 120, 125
unreliability of 114
transparency 19, 20, 23, 36, 58, 70, 85, 156, 186, 187
transport infrastructure 3, 5, 8
TSO (transmission system operator) 29
Turkey 78

Ukraine 78
uncertainty *see* risk
UNEP (United Nations Environment Programme) 4
on green jobs 6

United Nations (UN) 17–18
United States 78, 93–96, 149
  barriers to renewables in 130–131, 134–135, 136, 186
  Energy Dept. 4
  energy market operators in 158
  Energy Policy Act (1992) 135
  federal action in 95–96
  fixed-price incentives in 93
  green economy in 2–4, 9
  Green Jobs Act (2009) 2–4
  green jobs in 7
  green power programmes in 161–164
  grid integration in 124–125, 126
  net metering in 164, 165, 166
  nuclear power in 113, 137–138, 141, 142, 144
  political factors in 134–135, 189–190
  power outages in 112–113
  PURPA Act (1978) 59, 77, 93
  R&D programmes in 167
  Recovery and Reinvestment Act (2009) 2, 3
  RECs in 158
  RPS in 151, 153–154
  SBCs in 167–169
  SRECs in 10
  state schemes in 93–95
  storage technology in 121–123
  tax credits in 172, 173, 174, 175
  wind power in 116–117, 135, 141, 142, 144, 145, 155, 174
  see also specific states

variability *see* reliability
vegetable oil 119
vehicles, electric/hybrid 3, 121
Vermont (US) 90, 93, 94, 95
Victoria (Australia) 100–101
Vine, Edward 112
Virginia (US) 94, 166
visual impact issues 11, 36, 55, 145
voltage dips 46

Washington State (US) 93, 94, 113–114, 151, 169
waste 8, 26, 89, 118
  and FITs 17
  see also landfill gas
water efficiency 8
wave energy 86
Western Australia 101–102
wildlife 140–141, 142, 143
windfall profits 15, 39, 47, 59, 62
wind maps 16
wind power/farms 9, 18, 43, 59, 79, 86, 92, 106, 107, 153
  and avian mortality 140–142, 143
  in combined technology/hybrid schemes 44, 116–118
  and community partnerships 10–11, 42
  external costs of 140–142, 143, 144
  FIT calculations for 21, 22, 25, 28, 63, 105
  and forecasts of production 44–45
  and grid integration 124–125
  increasing tariffs for 52
  lead times for 34
  location of 33, 47, 52, 145
  and mini-grids 73
  offshore turbines 17–18, 33, 49, 82
  opposition to 11, 145
  potential of 4–5, 7
  premium FIT and 40
  reliability of 114, 116–118
  repowering 53–54
  scale/size issues with 69
  and solar plants 116–118
  and spatial planning 35
  tariff degression for 49
  variability of 30, 41, 44, 112
  and voltage dips 46
wind power studies 116–117
Wisconsin (US) 93, 116–117, 151
World Wind Energy Association 91

Zambia 119